SAE EDGE™
RESEARCH REPORT

Unsettled Topics in Unmanned Aerial Vehicle Icing

Richard Hann
Norwegian University of Science and Technology (NTNU)
Tor Arne Johansen
Norwegian University of Science and Technology (NTNU)

EDGE DEVELOPMENT TEAM

Ali Benmeddour, *National Research Council Canada*
Ben Bernstein, *Leading Edge Atmospherics*
Martin Fengler, *Meteomatics*
David A. Funkhouser, *Boeing Insitu*
Morten Hansbø, *Norwegian Defence Research Establishment*
Torbjørn Houge, *Maritime Robotics*
Philip McGillivary, *United States Coast Guard*
Riku Niemenmaa, *Finnish Defence Forces*
Isik Ozcer, *ANSYS*
Paul Pellicano, *Federal Aviation Administration*
Dennis Regnier, *CAV Systems*
Kim Lynge Sørensen, *UBIQ Aerospace*
Rune Storvold, *NORCE Norwegian Research Centre*
Kevin Yugulis, *Battelle Memorial Institute*

SAE INTERNATIONAL®
Warrendale, Pennsylvania, USA

About the Publisher

SAE International® is a global association of more than 128,000 engineers and related technical experts in the aerospace, automotive, and commercial-vehicle industries. Our core competencies are lifelong learning and voluntary consensus standards development. Visit sae.org

SAE EDGE™ Research Report Disclaimer

SAE EDGE™ Research Reports focus on topics that are dynamic, in which knowledge is incomplete, and which have yet to be standardized. They represent the collective wisdom of a group of experts and serve as a practical guide to the reader in understanding unsettled subject matter. They are not meant to provide a recommended practice or protocol. The experts have assembled as a community of practitioners to contribute and collectivize their thoughts and points of view; these are not the positions of the institutions or businesses with which they are affiliated nor are one contributor's perspective advanced over other contributors. SAE EDGE™ Research Reports are the property of SAE International and SAE alone is responsible for their content.

About This Publication

SAE EDGE™ Research Reports provide state-of-the-art and state-of-the-industry examinations of the most significant topics in mobility engineering. SAE EDGE™ contributors are experts from research, academia, and industry who have come together to explore and define the most critical advancements, challenges, and future direction in areas such as vehicle automation, unmanned aircraft, cybersecurity, advanced propulsion, advanced manufacturing, Internet of Things, and connectivity.

Related Resources

SAE MOBILUS® Automated & Connected Knowledge Hub
https://saemobilus.sae.org/automated-connected

SAE Team

Frank Menchaca, Chief Product Officer
Michael Thompson, Director, Standards, Information and Research Publications
Monica Nogueira, Acquisitions Director
Beth Ellen Dibeler, Product Manager
William Kucinski, Managing Technical Editor

Printed in USA

EPR2020008
ISSN 2640-3536
e-ISSN 2640-3544
ISBN 978-1-4686-0169-5

To purchase bulk quantities, please contact: SAE Customer Service

E-mail: CustomerService@sae.org
Phone: 877-606-7323 (*inside USA and Canada*)
+1-724-776-4970 (*outside USA*)
Fax: +1-724-776-0790

https://www.sae.org/publications/edge-research-reports

About the Editors

Richard Hann is a researcher at the Norwegian University of Science and Technology (NTNU) on the topic of icing on unmanned aerial vehicles. In 2013, he graduated with excellence from the University of Stuttgart in Germany as an aerospace engineer. After working for three years in the petroleum industry as an upstream project engineer, he returned to academia in 2016 to start a PhD on icing of unmanned aircraft. He is part of the Centre for Autonomous Marine Operations and Systems (NTNU-AMOS) and the Centre for Integrated Remote Sensing and Forecasting for Arctic Operations (CIRFA). Richard has more than eight years of experience with numerical and experimental icing aerodynamics on wind turbines and aircraft. Today, he is one of the leading researchers in the emerging research field of icing on unmanned aircraft. He holds a position as lead aerodynamics engineer and shareholder at UBIQ Aerospace. Richard is also promoting the application of drone technology in the Arctic with several ongoing projects in the fields of meteorology, glaciology, and atmospheric pollution at the University Centre in Svalbard.

Tor Arne Johansen received his MSc in 1989 and PhD in 1994, both in electrical and computer engineering, from NTNU in Trondheim, Norway. From 1995 to 1997, he worked at SINTEF as a researcher before he was appointed Associated Professor at NTNU in 1997 and Professor in 2001. He has published several hundred articles in the areas of control, estimation, and optimization with applications in the marine, aerospace, automotive, biomedical, and process industries. In 2002, he co-founded the company Marine Cybernetics AS where he was Vice President until 2008. Prof Johansen received the 2006 Arch T. Colwell Merit Award from SAE International. He is currently a principal researcher with NTNU-AMOS and director of the Unmanned Aerial Vehicle Laboratory at NTNU and the SmallSat Laboratory at NTNU. He recently co-founded the spin-off companies Scout Drone Inspection, UBIQ Aerospace, and Zeabuz.

contents

Atosan/Shutterstock.com

Unsettled Topics in Unmanned Aerial Vehicle Icing

Abstract

Unmanned aerial vehicles (UAVs) are an emerging technology with a large variety of commercial and military applications. In-flight icing occurs during flight in supercooled clouds or freezing precipitation and is a potential hazard to all aircraft. In-flight icing on UAVs imposes a major limitation on the operational envelope. This report describes the unsettled topics related to UAV icing. First, typical UAV applications and the general hazards of icing are described. Second, an overview of the special technical characteristics of icing on autonomous and unmanned aircraft is given. Third, the operational challenges for flight in icing conditions are discussed. Fourth, technologies for ice protection that mitigate the icing hazard are introduced. Fifth, the tools and methods required to understand UAV icing and to develop aircraft with cold-weather capabilities are presented. Finally, an assessment of the current and future regulations regarding icing on UAVs is provided.

Icing is a key challenge that the UAV industry needs to address in order to unlock the full potential of this emerging technology. UAVs must be capable of safe and reliable operation in a wide range of weather conditions. This report outlines the most important challenges and gives short- and long-term recommendations on how to solve UAV icing issues.

NOTE: SAE EDGE™ Research Reports are intended to identify and illuminate key issues in emerging, but still unsettled, technologies of interest to the mobility industry. The goal of SAE EDGE™ Research Reports is to stimulate discussion and work in the hope of promoting and speeding resolution of identified issues. SAE EDGE™ Research Reports are not intended to resolve the challenges they identify or close any topic to further scrutiny.

RICHARD HANN
Norwegian University of Science and Technology (NTNU)

TOR ARNE JOHANSEN
Norwegian University of Science and Technology (NTNU)

Edge Development Team

Ali Benmeddour, *National Research Council Canada*
Ben Bernstein, *Leading Edge Atmospherics*
Martin Fengler, *Meteomatics*
David A. Funkhouser, *Boeing Insitu*
Morten Hansbø, *Norwegian Defence Research Establishment*
Torbjørn Houge, *Maritime Robotics*
Philip McGillivary, *United States Coast Guard*
Riku Niemenmaa, *Finnish Defence Forces*
Isik Ozcer, *ANSYS*
Paul Pellicano, *Federal Aviation Administration*
Dennis Regnier, *CAV Systems*
Kim Lynge Sørensen, *UBIQ Aerospace*
Rune Storvold, *NORCE Norwegian Research Centre*
Kevin Yugulis, *Battelle Memorial Institute*

ISSN 2640-3536

Introduction

When people think about drones, they usually picture the large military unmanned aircraft of controversial newspaper headlines or they think about the noisy toy quadcopters the neighbor flies in the backyard. In truth, there is a wide range of unmanned aerial vehicles (UAVs) between these two extremes. For example, UAVs are used for delivering urgent medical supplies, providing broadband access to remote areas, performing search and rescue missions, and exploring scientific research questions. UAVs are becoming more and more a part of our everyday lives - a trend that is going to increase in the near future. Companies are now testing UAVs that can deliver packages to our doorsteps, minutes after the products have been purchased online. New disruptive businesses are also exploring the possibilities of designing unmanned air taxis to transport passengers (urban air mobility).

Many of these proposed UAV applications rely on the ability of the aircraft to reach areas that are difficult for humans to access. Examples include places that are either far away, lack infrastructure, or are hazardous to humans (in the case of dull, dirty, and dangerous jobs). UAVs themselves may become a hazard to the public, when operating above populated areas. Therefore, it can be very important that a UAV is able to perform its mission in nearly all weather conditions. One particular weather phenomenon that is a danger to UAVs is atmospheric in-flight icing. Atmospheric icing is a well-known topic to the manned aviation community but is a relatively new and emerging topic for UAVs. The objective of this report is to present an overview of the unsettled topics related to UAV icing, and to identify key research and technology gaps that need to be addressed by the industry to fully exploit the potential of UAVs.

UAV Applications

The first UAVs were developed and used in the early 1900s for military applications. Since then, UAV technologies have further developed and become an integral part of most modern defense forces around the world. Military UAVs come in all types and sizes, ranging from hand-launched micro UAVs to large, high-altitude UAVs that are comparable in size to an airliner. The most common military UAV applications are related to intelligence, reconnaissance, surveillance, security, attack, combat support, and sustainment, as well as command, control, and communication support. These activities are of high importance and directly contribute to mission success. Consequently, UAVs need to be able to carry out these tasks independently from weather limitations, including atmospheric icing.

Parallel to military applications, recreational use of UAVs also has a long history. Radio-controlled model airplanes have been the focus of hobbyists for years, and quadcopters have recently become a widely available consumer product. As UAV technologies became more affordable and accessible, commercial use has increased as well. Today, UAVs are utilized for a wide range of civil applications [1].

One key differentiation of UAV applications is between operation within visual line of sight (VLOS) and beyond visual line of sight (BVLOS). Today, most commercial UAV operations are conducted in VLOS with rotary wing UAVs and limited automation and autonomy. UAVs used for these applications are easy to operate and are mostly used to for "birds-eye view" imagery. Currently, the largest users of these UAVs are in the construction and agriculture sectors, where UAVs are used for surveying, stockpile volume measurements, and crop monitoring. Notably, smaller scale UAV applications include photography, film, inspection, mapping, remote operations, research, and search and rescue.

There are many ideas for new UAV applications that rely on BLVOS operations with either remotely piloted aircraft systems (RPAS) or completely autonomous aircraft. Such UAVs are typically fixed-wing and often have automated and/or vertical takeoff and landing capabilities. The most prominent and recent example of this would be UAV package delivery services. Another emerging sector is the energy industry, where UAVs can be used for monitoring power lines, pipelines, storage tanks, solar panels, oil spills, and more.

One very important future domain for UAVs is the Arctic. Climate change is occurring at much faster rates in the Arctic than compared to the rest of the world - making it a highly relevant research object for climate scientists [2]. Today, the Arctic is also a focus point for geopolitics and security issues. Last but not least, the opening of the Northwest Passage between Europe and Asia offers important future economic opportunities in the shipping sector. Since the Arctic is a very remote area with limited satellite coverage, UAV technology is a key element for many future operations - commercial, scientific, or military. One example is ship-launched UAVs that can detect and track icebergs and sea ice, offering increased navigation safety for shipping.

Table 1 gives an overview of existing and new civil applications for UAVs and the operational range (VLOS or BVLOS) associated to them. Technically, all UAVs that operate BVLOS are exposed to the risk of icing and require a risk-mitigation strategy. Depending on the application, mitigation

TABLE 1. VLOS or BVLOS operation of commercial UAVs by application.

Civil UAV applications	VLOS	BVLOS
Photography and filming	×	×
Inspection and monitoring	×	×
Mapping and surveying	×	×
Research	×	×
Search and rescue	×	×
Emergency response	×	×
Meteorology	×	×
Communications	×	×
Remote operations		×
Deliveries		×
Arctic operations		×

Foreground: © Richard Hann. Background: Dmitry Kalinovsky/Shutterstock.com solutions may consist of using nowcasting and forecasting data to avoid ice conditions, or dedicated ice protection systems (IPS) that allow the aircraft to operate safely within icing conditions for a limited period of time. VLOS operations are typically less susceptible to in-cloud icing but may still be affected by freezing precipitation events.

State of the Industry

Market forecasts on the expected growth of the global UAV market vary in numbers, but generally agree that the market will see a compound annual growth rate (CAGR) of 10 to 20 percent during the next years (Figure 1). The market can be divided into military and commercial segments; historically, the military has been the largest operator of UAVs, and this is reflected in a significantly larger market share today [3]. However, the commercial UAV market has seen a rapid growth in recent years [4], which is expected to further increase in the future. A recent report from the Federal Aviation Administration (FAA) forecasts that the commercial market in the United States alone is likely to triple between 2019 and 2023 [5].

In-Flight Icing

In-flight icing, also called atmospheric icing, occurs when an aircraft encounters supercooled liquid water in the atmosphere and that liquid freezes onto the aircraft. The water occurs as cloud droplets, or precipitation in liquid form, with a temperature below the freezing point. When such supercooled droplets collide with an aircraft, they freeze on the surface and can grow into various ice shapes. Atmospheric icing conditions can be encountered around the globe any time of year [6, 7]. The ice accretion rate is dominated by the air temperature, size, and velocity of the vehicle, but also by the liquid water content (LWC) and the droplet size, often expressed in terms of median volume diameter (MVD) [8]. The resulting ice shapes can be categorized into several icing morphologies. The counterpart of in-flight icing is ground icing, during which aircraft accumulate ice prior to takeoff due to supercooled fog, frost, or precipitation.

Ice Types

Rime Ice Rime ice typically occurs at low temperatures, when all impinging droplets freeze immediately upon impact on a cold surface. During the ice accretion process, small pockets of air are enclosed in between the freezing droplets. Consequently, rime ice appears to be white and exhibits a rough surface with small protruding ice feathers (Figure 2). Rime ice shapes typically have a streamlined form with limited effect on the airfoil aerodynamics except for cases with extensive icing durations [9].

Glaze Ice When temperatures are close to the freezing point, not all droplets freeze instantly upon impact on the airframe. Consequently, a thin, liquid water film will develop on the surface that will gradually freeze. The resulting ice form is called glaze or clear ice, as it appears translucent due to the lack of air inclusions (Figure 2). Glaze icing often builds into irregular ice shapes that can grow into large ice horns with high aerodynamic penalties [9].

Mixed Ice In practice, ice often occurs as a combination of glaze and rime ice, called mixed ice. Mixed ice is characterized by partial freezing of impinging droplets and the simultaneous formation of a liquid water film on the surface. The ice shapes come in a large variety of forms that may result in horn-like structures. These ice horns can result in significant aerodynamic performance degradations [10].

FIGURE 1. Global UAV market size and forecast 2018-2025 [3, 4].

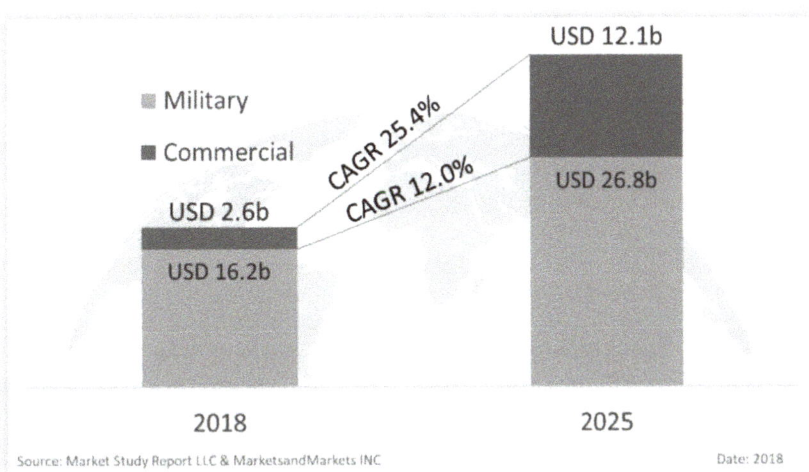

Foreground: © Richard Hann. Background: Intellson/ Shutterstock.com

FIGURE 2. Rime ice (left) and glaze ice (right) accretions on a UAV airfoil from an icing wind-tunnel test.

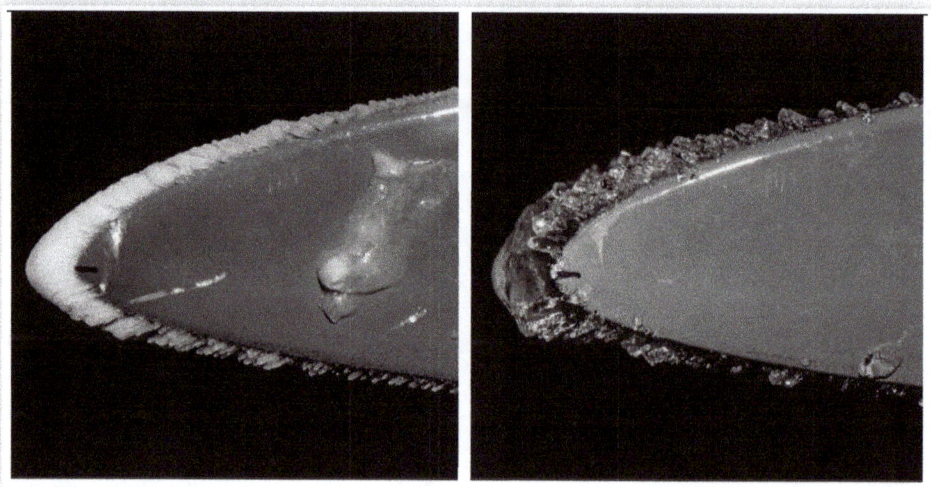

FIGURE 3. Trajectories of small and large droplets. Droplet size is influencing the collision efficiency.

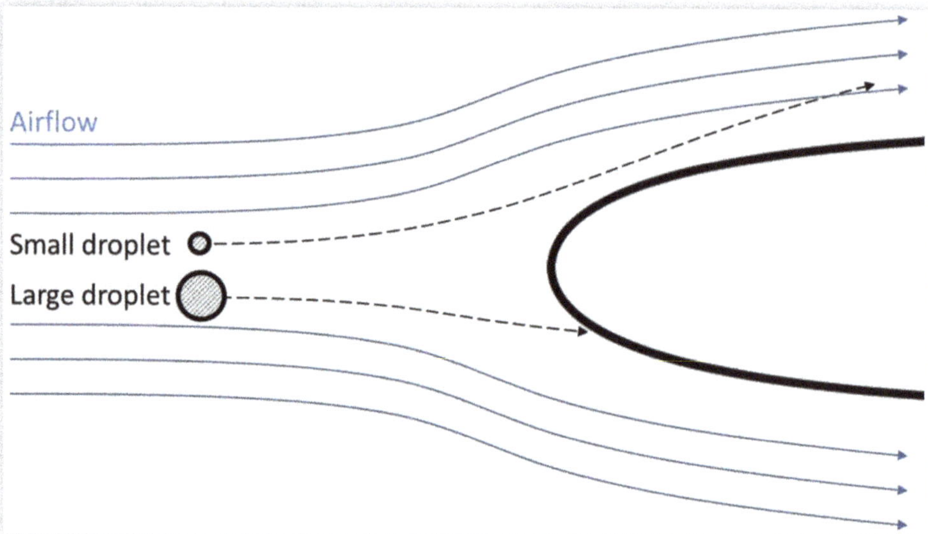

Supercooled Large Droplets Most icing events occur in clouds (also referred to as in-cloud icing) with supercooled cloud droplets with MVDs below 40 to 50 microns. However, there is another important icing regime that occurs during precipitation with supercooled droplets. This regime is characterized by environments with MVDs exceeding 40 to 50 microns, with values that can be as high as several thousand microns. Therefore, these droplets are also called supercooled large droplets (SLDs) which can occur during freezing drizzle and freezing rain events [11]. SLD icing conditions can lead to extremely high ice-accretion rates that may cover a significantly larger surface area of the lifting surface compared to in-cloud icing, and is thus considered particularly hazardous [9].

Snow and Ice Crystals Snow is generally a lesser threat to manned aircraft, as snow crystals typically do not stick to the lifting surfaces due to the high flight velocities. Ingestion of large amount of ice crystals can however lead to adverse effects on the engine [12]. Snow is more a relevant form of icing for static structures on the ground, such as power lines, communication towers, meteorological masts, and wind turbines. Wet snow, in particular, can lead to high snow-accumulation rates on structures that can result in excessive weight loads and, consequently, mechanical failures [13]. However, snow may be an issue for rotary wing UAVs or airships that are stationary for extended periods in wet snow clouds (Figure 8).

Cold Soaking The cold-soak effect occurs when aircraft that carry very cold fuel in their wings encounter precipitation. In this case, the cold fuel cools the wings below the freezing point, so that precipitation will freeze as clear ice on the wings, even if the local air temperature is well above freezing. This icing mechanism can be relevant especially for high-altitude long-endurance (HALE) UAVs.

Icing Effects

Numerous wind-tunnel experiments, in-flight tests, and numerical simulations show that ice accumulated on the leading edge of a lifting surface will lead to a degradation of its aerodynamic performance [14]. The ice shapes modify the airfoil geometry and typically lead to a significant decrease in lift, increase in drag, change in pitch moments, and deterioration of the stall behavior (Figure 4) [8, 9]. Furthermore, icing negatively affects aircraft stability and control [15]. The degree of performance degradation depends on the form of the ice shapes and the degree of disruption of the airflow. A numerical study on the icing penalties of a typical UAV airfoil for a wide range of meteorological icing parameters showed that lift can be decreased by 35 percent, stall angles can be reduced by 33 percent, and drag can be increased by up to 400 percent in the linear region [16].

In general, four different ice morphologies can be identified [9]: ice roughness, horn ice, streamwise ice, and spanwise-ridge ice (Figure 5). In the beginning of the icing process - before larger horns develop - ice forms a rough surface layer. This surface roughness increases skin friction and can trigger early laminar-turbulent transitioning of the boundary layer. These effects lead to an increase in drag and a reduction of the stall angle. Horn ice is a complex ice shape that is often formed by glaze and mixed ice conditions. This ice is characterized by protruding horn features that lead to a flow separation from the top of the horns. The resulting leading-edge flow separation triggers an early

FIGURE 4. Example of the aerodynamic performance degradation due to icing on a UAV airfoil.

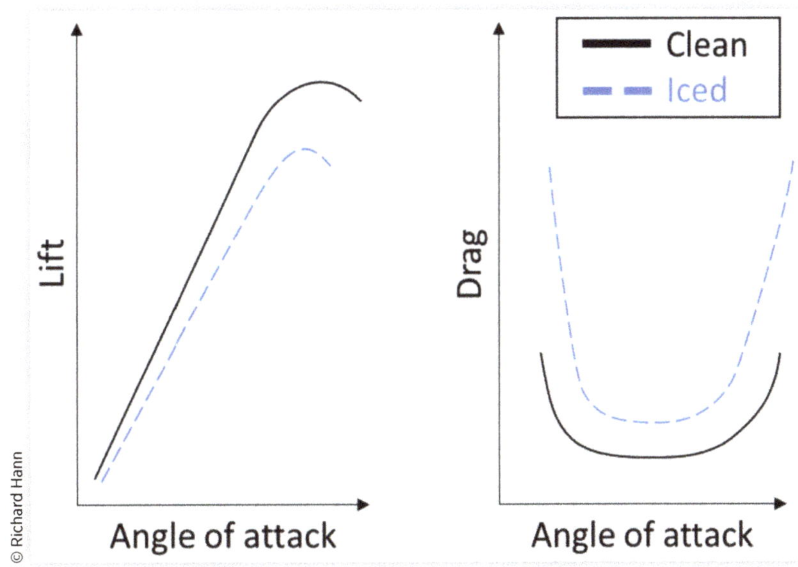

FIGURE 5. Typical ice morphologies on an airfoil.

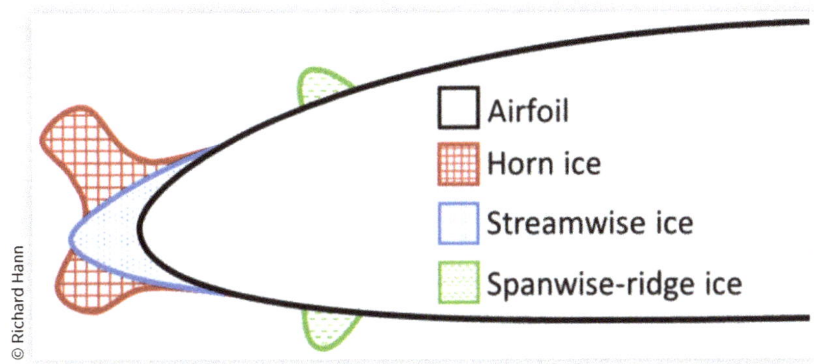

laminar-turbulent transition, significantly increases pressure drag, and reduces lift. Streamwise ice is typically formed during rime ice conditions and results in streamlined ice shapes. The effect on the flow field is much smaller compared to horn ice because leading-edge separations are small or absent. Spanwise-ridge ice occurs in combination with IPS that only protects a limited area. Spanwise ice ridges can form when droplets impinge behind the protected area (e.g., during SLD icing) and/or when the protected surface is heated and the resulting runback liquid water film refreezes in an unprotected area of the airfoil (runback icing). The resulting ice shapes act as a spanwise flow obstacle that can have significant effects on the performance through early transition and flow separation effects. In terms of aerodynamic degradation severity, spanwise-ridge ice has the highest impact, followed by horn ice, streamwise ice, and ice roughness [9]. SLD and runback icing may also extend to control surfaces, decreasing their effectiveness or blocking them.

Icing in Manned Aviation

The history of icing studies on manned aircraft dates back to the 1940s and 1950s, when the foundation for modern icing research was laid [8]. Many experiments and flights tests have been performed to develop a deeper understanding of ice-accretion processes and to design the first IPS [17]. The advent of computers in the 1970s and 1980s opened the door for developing sophisticated icing models that could be solved numerically. This has led to the development of the first generation of icing analysis tools, some of which are still in use today. During the 1990s the topic of SLDs icing became prominent, especially after the crash of American Eagle Flight 4184, leading to an increased awareness of the hazards of freezing precipitation [18].

Today, icing in manned aviation is generally a well-understood problem. A large amount of research has been performed to understand the consequences of icing on aircraft systems such as lifting surfaces [8, 9, 15], propellers 19], rotors [20, 21], pitot tubes [22], carburetor [23], engines [12], and inlets [24]. Furthermore, aircraft icing and mitigation techniques are addressed during aircraft certification [25] and pilot education [26]. Nonetheless, icing research remains an open field. Recent developments aim to develop better numerical simulation tools, improve real-time nowcasting and forecasting, and introduce new technologies for ice detection and ice mitigation.

Icing on UAVs

The first mention of UAV icing in the open literature dates back to 1990 in a comprehensive study by the United States Naval Air Development Center, describing the hazards of icing for military UAV operations [27]. Further reports state that icing was responsible for UAV crashes in Hungary, Afghanistan, Serbia, and Kosovo during the 1990s [28, 29]. Since then, very little information is openly available, which is likely related to the fact that - up until recently - most UAV operations were performed by the military. One more recent incident, which became publicly known, happened in February 2017, when a British Army Watchkeeper UAV stalled after its pitot tube got blocked, most likely due to icing [30].

Part of the reason why UAV icing has not been addressed earlier is that most military UAV applications were designed for operation in hot weather environments. Another barrier to the development of all-weather capable UAVs is the high cost for technology development, since no mature solutions are available yet. Since UAVs are now becoming more and more viable for commercial applications, the UAV icing topic is shifting more into focus. During the last 10 years, an increasing number of publications have become available on the subject. The research has been focused on aerodynamic performance degradation [16, 31, 32, 33, 34], rotor icing [35, 36], IPS [37, 38], ice detection [39, 40], and path planning [41]. However, most of the research is only scratching the surface of the topic, and an in-depth understanding of UAV icing has not been achieved yet. This report aims to identify the gaps in research and the unsettled technical problems.

Manned versus Unmanned

Commercial and military all-weather capable UAVs are expected to be just as reliable as piloted aircraft. Therefore, the icing issue needs to be solved just the same, but there are several differences between manned and unmanned aircraft that are relevant in that context. The comparison between the two types of aircraft is somewhat difficult, since both come in a large variety of forms and sizes. Since a considerable amount of icing research has been performed on large passenger transport airplanes, these will be considered as a "typical" manned aircraft for the purpose of this report. These aircraft typically travel at similar speeds but can vary greatly in size and weight. UAVs come at an even greater variation of forms and sizes, ranging from hand-launched micro UAVs to large, high-altitude military aircraft (Figure 6). Table 2 shows the specifications of selected UAVs and compares them to a small and large manned passenger transport aircraft.

Generally, unmanned aircraft tend to fly at lower velocities compared to manned aircraft. The reason for this is that most UAV mission profiles are endurance driven with the objective to loiter for extended durations above an area of interest. Due to the lower speed requirements, most UAVs utilize propellers for propulsion, with electrical, piston, or turbo engines instead of jet engines. UAVs tend to be significantly lighter than manned aircraft as their payload capacity is smaller too. The wingspans of the largest UAVs are comparable to small manned passenger aircraft - but the majority of UAVs have much shorter wings. The operational altitude also varies for UAVs. On one end of the spectrum are large UAVs, used primarily for surveillance, which operate at altitudes higher than manned aircraft (e.g., HALE UAVs). On the other end are small UAVs that operate in limited areas, flying in close proximity to the ground

FIGURE 6. **Comparison of different military UAV sizes.**

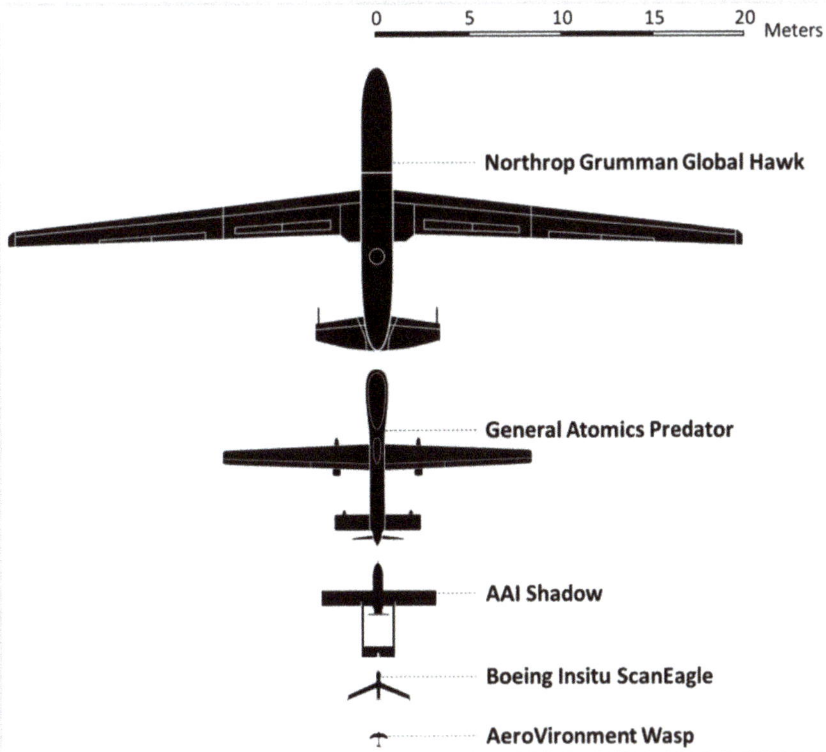

level. Finally, the most obvious difference is that UAVs do not have a pilot on board that can identify icing conditions; instead, they must rely completely on onboard instruments. Consequently, the overall degree of automation tends to be larger in UAVs. It should be noted that icing on manned rotorcraft has several similarities to UAVs, as helicopters are smaller in size (especially the lifting surfaces) and operate closer to the ground. Furthermore, larger UAVs share similarities with general aviation, in particular with regard to size and velocity.

These differences can have significant impact on icing processes and have effects that have not been fully understood or investigated yet. The next sections will discuss the main mechanics of UAV icing and the different icing behaviors between manned and unmanned aircraft.

Wind Energy

On a side note, it should be mentioned that UAV icing shares - perhaps surprisingly - many aspects with icing on wind turbines. Atmospheric icing is a well-known challenge for wind energy in cold climate conditions. Icing on wind turbines can lead to significant performance losses, structural damages to the blades, hazard of ice throw, and an increase in noise levels [42].

Compared to icing in aviation, wind turbine icing is a relatively new topic. Research and the development of mitigation technologies started in the 1990s. In the recent years, the technological solutions have matured significantly, and today most wind turbine manufacturers offer off-the-shelf IPS for their turbine blades [43].

TABLE 2. **Comparison of typical UAV and large transport aircraft characteristics.**

	Span	MTOW	Cruise	Ceiling
Northrop Grumman Global Hawk RQ-4B	39.9 m	11,600 kg	160 m/s	60,000 ft
General Atomics Preator MQ-1B	14.8 m	1,000 kg	41 m/s	25,000 ft
AAI Shadow RQ-7B V2	6.2 m	212 kg	38 m/s	18,000 ft
Boeing Insitu ScanEagle	3.1 m	22 kg	31 m/s	19,500 ft
AeroVironment Wasp	1.0 m	1.3 kg	10 m/s	500 ft AGL
Boeing 737 MAX 8	35.9 m	82,200 kg	233 m/s	41,000 ft
Airbus A-380-800	79.8 m	575,000 kg	250 m/s	43,000 ft

FIGURE 7. Icing on the rotor of a quadcopter.

There are several similarities between UAVs and wind turbines that may offer synergies to both applications. First, both technologies must deal autonomously with icing risks. This requires particularly robust ice-detection systems with high reliability, along with well-tuned control algorithms. The barrier to test new ice-detection technologies on wind turbines is much lower than in aviation, due to more relaxed certification constraints. This has led to the development of a wide variety of ice-detection systems that are currently used in wind energy, some of which are low cost and low weight and may therefore be well suited for UAV applications. Innovative detection systems that are in use on wind turbines today utilize ultrasonic waves, capacitance change, impedance change, light reflectance, microwaves, change in eigenfrequencies, operational modal analysis, and more [44]. Second, icing on wind turbines and UAVs occur typically at Reynolds numbers an order of magnitude lower than those encountered in manned aviation. Third, both UAVs and wind turbines encounter in-cloud icing at significantly lower velocities and altitudes compared to aviation, and are potentially exposed for longer durations. Wind turbine icing events can last several hours or even days, leading to large ice accretions and can be compared to, for example, a surveillance UAV loitering in icing conditions.

Technical Characteristics

UAVs are different from manned aircraft in several key aspects. These characteristics result in special behavior when it comes to icing. This section explores the most important technical differences between UAVs and manned aircraft and explains how these differences interact with icing.

Vehicle Type

The effects and severity of icing on a UAV inherently depend on the vehicle type. Icing on rotary-wing UAVs will mainly affect the rotors (Figure 7). UAV rotor blades are typically small and spin fast which makes them accrete ice rapidly. Ice accretions on the rotors lead to the loss of uplifting force, introduce imbalances, and increase drag of the rotor blades, requiring additional power to maintain rotational speed. This can quickly lead to the loss of control of the aircraft and, in some cases, crashes.

Typically, the high ice-sensitivity of rotors leaves airframe icing on rotary-wing UAVs a secondary importance - although it should be noted that the downwash of the rotors may increase icing rates locally. Snow usually does not accumulate on fixed-wing aircrafts, but it may accumulate on a rotorcraft that is hovering for extended periods of time in snow conditions (Figure 8).

On fixed-wing UAVs, the ice accretion on wings and propellers is critical. Ice on the wings changes the geometry of the airfoils and leads to a degradation of the aerodynamic performance. Icing increases drag and decreases lift of the wings, requiring the aircraft to compensate with increased thrust and/or higher angles of attack. Ice also affects the pitching moment generated by the wing and the location of the center of gravity, which may significantly reduce the stability margins. Flow separation (stall) is likely to occur earlier and more abruptly on an iced wing. Combined, these effects result in a high risk of the aircraft losing control or crashing during icing conditions. Engine icing may also be a hazard, when the engine ingests large amounts of snow or ice, or when carburetor icing occurs in engines without fuel injection. Icing or snow may block cooling inlets necessary for the operation of electric propulsion systems. Since smaller UAVs typically do not have propulsion system redundancy,

FIGURE 8. Snow accumulation on a quadcopter.

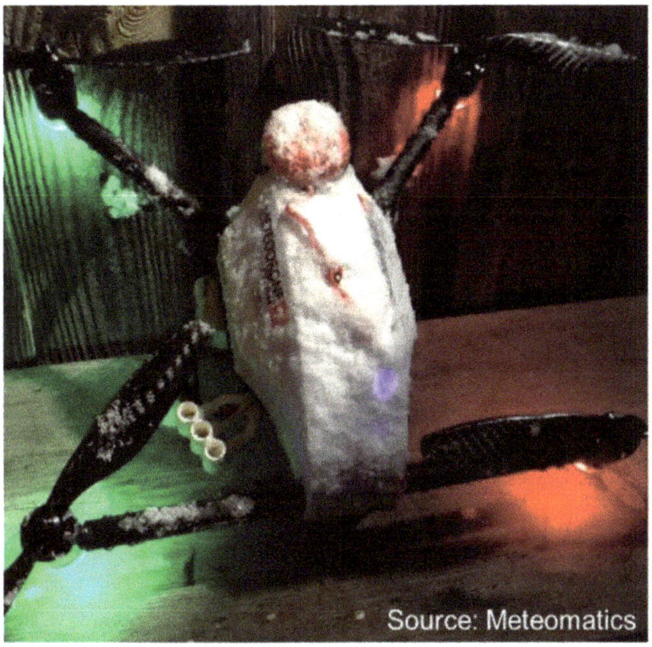

a single engine failure is especially hazardous. Last but not least, small UAVs may have external control surface actuators that can be prone to icing, leading to a loss of maneuverability.

Size

UAVs come in a large variety of sizes with wingspans ranging from centimeters to tens of meters (Figure 6 and Table 2). Each UAV type has its own individual icing challenges. Compared to manned aircraft, UAVs are generally significantly smaller, except for the largest military UAVs. One effect of this size difference is that UAV airfoils tend to build larger ice horns relative to their size, compared to manned aircraft [8]. Figure 9a shows numerical simulation results for the ice accretion on a NACA0012 airfoil for different chord lengths, at the same flight velocity. The airfoil with the smallest chord results with the largest ice thickness relative to its chord length (relative ice thickness) and the largest relative area covered by ice [45, 46].

FIGURE 9. Ice shapes for different chord lengths on a NACA0012 airfoil, simulated with LEWICE. Normalized with chord length (left) and in absolute units (right).

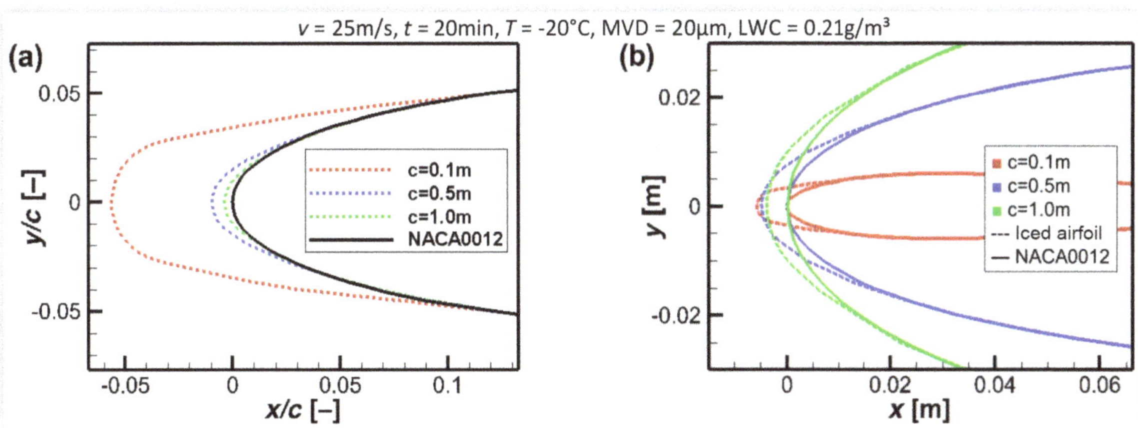

This can be explained with droplet trajectory physics. The trajectory of a droplet around an airfoil is governed by droplet inertia and the aerodynamic forces acting on the droplet (Figure 3). The trajectory of a small droplet is dominated by aerodynamic forces and tends to follow streamlines. The trajectory of a large droplet is dominated by inertia and follows straight-lined trajectories that are more likely to collide with an airfoil. The magnitude of the aerodynamic forces responsible for deflecting droplets forming on the airfoil is governed by the size and form of the airfoil. Small and thin airfoils have a lower effect on the flow field, and generate weaker aerodynamic forces compared to large and thick objects. For smaller airfoils, droplets will be less likely to follow the streamlines, but instead travel more in a straight line and be more likely to hit the airfoil (i.e., resulting in a higher collision efficiency). Consequently, the smaller airfoil will accumulate ice over a larger relative area, with higher relative ice thickness, while the total ice masses will be lower compared to a larger airfoil (Figure 9b). Typically, the relative ice shape thickness can be directly correlated with the aerodynamic performance degradation [9].

In summary, smaller aircraft tend to accumulate thicker ice shapes relative to their size, which result in larger aerodynamic performance penalties compared to a larger aircraft. In other words, UAVs tend to be more sensitive to icing conditions compared to larger manned aircraft. This also implies that icing conditions that are considered light for manned aircraft may be more severe for UAVs [47].

Flight Velocity

Due to a smaller size, different mission profiles, and propulsion concepts, UAVs are typically moving at lower velocities than manned aircraft. Table 2 shows the large variation of flight speeds of selected UAVs [48]. At high velocities the effect of aerodynamic heating due to viscous friction and air compression leads to an increase in surface temperatures of the airframe. On high-speed manned aircraft, this effect is often large enough to prevent ice formation at air temperatures just below the freezing point, especially on propellers. Due to the lower flight velocities of UAVs, aerodynamic heating is generally a negligible effect. Furthermore, the lower suction pressure (resulting from lower air speeds) leads to reduced surface cooling and evaporation rates. Consequently, icing is less likely to occur at temperatures above freezing point [49]. Shear stresses promoting ice shedding are also reduced due to lower airspeed and friction [50]. Moreover, lower flight velocities lead to a decrease of ice accretion rates. However, this effect is partially counterbalanced by the increased exposure time to the icing conditions (i.e., ice accumulates slower on a UAV, but it takes longer to fly through icing conditions). Furthermore, lower flight velocities typically lead to smaller droplet collision efficiency and smaller impingement areas [8]. However, as can be seen from Figure 10, the effect of velocity tends to be smaller compared to the effect of size in Figure 9.

FIGURE 10. **Ice shapes for different flight velocities through an icing cloud with a horizontal extent of 17.4 nautical miles on a NACA0012 airfoil, simulated with LEWICE. Normalized with chord length.**

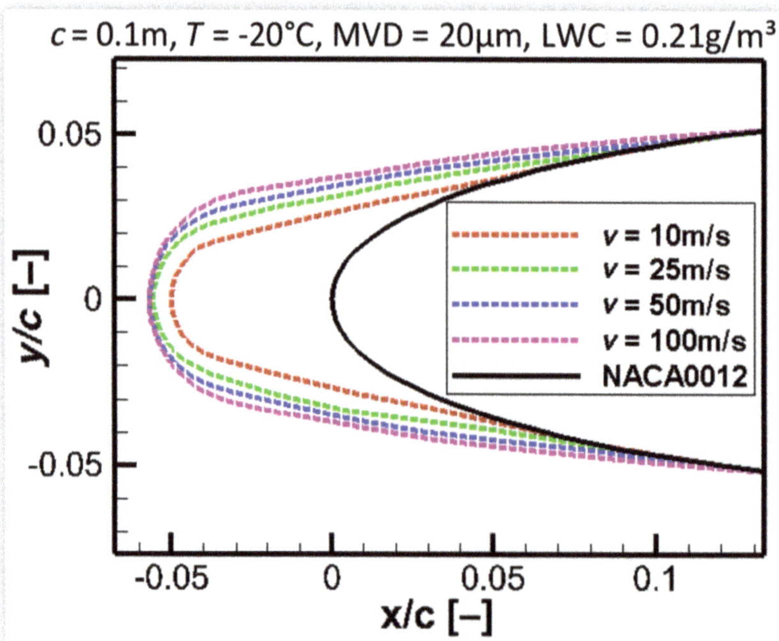

Reynolds Number

The Reynolds number is a dimensionless number characterizing the ratio of viscosity to inertia (momentum) in the fluid. It is used to characterize the flow regime for airfoil and wing aerodynamics and is strongly correlated with turbulence and transition. The combination of lower flight velocities and smaller aircraft sizes results in significantly lower Reynolds numbers for UAVs compared to manned aircraft. Whereas manned aircraft typically operate at high Reynolds numbers Re = 10^7-10^9, most UAVs operate at low Reynolds numbers Re = 10^5-10^7. The difference in the Reynolds number regime changes the physical behavior of the flow [46]. In the low Reynolds number regime, laminar flow is prevalent. Laminar flow is characterized by low drag and low resilience to separation, which contrasts with turbulent flow at higher Reynolds numbers, that is, high drag and more resilience to separation. UAV designs typically use specialized airfoils that maximize the extent of laminar flow to generate low drag. By their nature, these airfoils are very sensitive to contamination of their surfaces. Ice surface roughness - developing during the initial stages of icing - can trip the laminar airflow and lead to substantial drag increases [9]. Lastly, the difference in the Reynolds number regime also results in different wing designs for UAVs, typically leading to larger aspect ratios and lower sweep angles compared to manned aircraft.

Weight

Weight is the natural enemy of any aircraft. For smaller UAVs, even minor amounts of additional weight are significant in relation to their total weight. This characteristic, in combination with the increased icing sensitivity of small aircraft discussed above, implies that the weight of ice accretions can add significant penalties to the UAV performance (e.g., preventing the UAV from reaching its maximum altitude). The increase in airframe weight must be compensated by additional lift - either by increasing the angle of attack or by increasing flight velocity. This will have adverse effects on endurance, range, and stall margins of the aircraft and can endanger mission success. Furthermore, the ice accretions can also change the center of gravity and the pitch moment characteristics of the lifting surfaces. This can negatively affect the stability and maneuverability of the aircraft. The high sensitivity to additional weights is also a challenge when it comes to the design of ice detection systems or IPS. These systems must be very lightweight in order to not limit the utility of the UAV.

Materials

Icing on manned aircraft mainly occurs on materials such as aluminum or aluminum alloys with thermal conductivities in the order of approximately 200 watts per meter-Kelvin or "k value." In contrast, UAV airframes are often built of polymer-based composites with significantly lower conductivities of a k value of approximately 0.2. Recent research suggests that this difference in thermal conductivities can affect the ice accretion process significantly [51]. Experiments have shown that icing on polymer-based airfoils experiences a lower dissipation of latent heat of fusion from the freezing droplets. This leads to more runback water (especially for glaze), increased ice coverage, more complex ice shapes, and potentially larger aerodynamic penalties.

Energy

Energy is a limited resource on UAVs, especially when it comes to small UAVs. Small UAVs either are typically battery powered or use a combustion engine - only large UAVs are powered by turbine engines (jet or turboprop). The available energy type plays an important role in the design of IPS. Aircraft with combustion engines may use hot exhaust gases for heating of the airframe. Aircraft with turbine engines can use bleed air from the compressor for piccolo tube IPS. Electrothermal IPS can require high currents, which the aircraft power systems might not be able to supply without modifications or auxiliary systems. In summary, icing will have a negative effect on the endurance of any UAV. Unprotected aircraft require more thrust to compensate for the increase in drag and reduction of lift. Aircraft with IPS will be heavier and require additional energy to mitigate icing.

Rotor and Propeller Icing

Propellers and rotors are rotating lifting surfaces generating thrust for fixed-wing aircraft and lift for rotary-wing aircraft. Almost all current fixed-wing UAV designs use propellers for thrust generation, except for a few large military UAVs. Icing on rotors and propellers is therefore a critical topic, especially as ice accretion rates on rotors tend to be higher than on static surfaces. This is due to the smaller leading-edge diameters of rotor blades, as well as high relative air speeds, especially near the tip. Experiments have shown that icing can substantially degrade the performance of rotors and propellers and lead to excessive vibrations. One study on a UAV propeller shows that glaze ice conditions can lead to a thrust reduction of 75 percent coinciding with a required power increase of 250 percent after only 100 seconds of exposure to moderate icing conditions [35]. Icing on rotary-wing UAV rotor blades has similar negative effects and can build up very quickly [36]. In addition, icing can introduce imbalances between the rotors, leading to control issues and loss of stability.

Centrifugal forces acting on the ice accretions lead to shedding, once a critical ice mass has been reached. Ice shedding mainly occurs on the outer part of the rotating blades, where the centrifugal forces are largest [50]. Shedding efficiency depends also on the type of icing and is reduced if ice accretes over a large surface area (e.g., in SLD conditions) [52]. Ice shedding temporarily decreases the ice load on the propeller, but the shedding event itself leads to rotor imbalances. Partial shedding of ice on a rotor blade can also lead to severe imbalances. Furthermore, the shed ice fragments can hit other aircraft components and cause substantial damage [53].

Compared to manned aviation, rotors on UAVs are typically smaller, both in diameter and airfoil thickness. As discussed earlier, small leading-edge diameters lead to increased icing sensitivity and larger impingement areas. In particular, ice may accumulate aft of the leading edge due the small size of the blades in comparison to the droplet sizes (Figure 7) [54]. Also, in manned aviation, rotors and propellers experience high stagnation temperatures on the outboard span for the rotor blade, leading to a significant (approximately 10 to 25 degrees Celsius) increase of local temperature [8]. As a result, icing near the freezing point is typically not an issue near the tip of such rotors, as the stagnation temperature provides sufficient heating to prevent ice formations - however, it can still be an issue near the blade root. This effect is substantially diminished for UAVs with small rotor diameters, making them more sensitive to icing near freezing temperatures.

Sensor and Antenna Icing

Atmospheric icing can also affect the functionality of sensors. The most important sensor in this respect is the pitot tube. The pitot tube provides the airspeed to the autopilot. Due to its small size, the pitot tube is highly susceptible to icing. Ice accretions can block the pitot tube and/or the static port, leading to erroneous airspeed indications. The indicated airspeeds can freeze, drop to zero, or gradually change. Autopilots are typically not capable of detecting erroneous airspeed readings and may initiate maneuverers that can potentially crash the vehicle (stall or nose dive). Icing can also affect other types of sensors such as cameras, antennas, radomes, etc. Ice can impair sensor performance and increase the weight and drag of a vehicle. Sensors and antennas that are exposed to the airflow and that are very small are especially at risk of accumulating large amounts of ice in short timeframes. Icing on these components can degrade their functionality by reducing signal strengths and leading to communication loss [55]. This is a risk that is particularly relevant for remotely controlled UAVs.

Autopilot and Controls

The current trend in UAVs is moving away from RPAS to fully autonomous systems. A key element in this development is the autopilot, which takes responsibility for flight controls, navigation, flight path, landing, etc. Today, both large and small UAVs rely on autopilot systems with varying degrees of autonomy [56]. Atmospheric icing introduces changes to the vehicle's flight performance, stability, and control - this can be interpreted as a perturbation or fault from the clean (un-iced) flight state [57]. The ability of the autopilot to identify and adapt to such changes is essential for reliable operations of UAVs in icing conditions [58]. The identification of icing conditions can be performed either with a dedicated ice-detection sensor or by fault detection of the autopilot itself [40]. Adaptation to identified icing conditions can include increasing the cruise speed, changing in-flight path planning, limiting the flight envelope, and control surface deflection to respect the reduced stall margins.

Operational Challenges

A key challenge for operating UAVs is related to understanding the weather conditions that the vehicle is encountering, specifically knowledge of when an aircraft is entering and leaving hazardous icing conditions. For VLOS operations, a pilot is near the aircraft and can assess weather conditions, including the icing risks. This is obviously a highly subjective (and possibly biased) evaluation dependent on pilot experience and knowledge.

For BVLOS operations, UAVs must rely on their onboard instrumentation to detect the presence of icing conditions or ice accretions on their airframe. For successful mission planning, reliable icing nowcasting and forecasting products are becoming a key element. This leads to the question about how meteorological icing conditions may differ between UAVs and manned aircraft, and how these can be identified.

Ice Detection

Ice detection is a key element for unmanned aircraft that are operating BVLOS and for systems with all-weather capabilities. For UAVs with IPS, ice detectors are required to activate the IPS when icing conditions are encountered, to assess the icing severity, and to deactivate the system when icing conditions are left. UAVs without IPS require ice detection to identify hazardous conditions early enough that the aircraft can attempt to avoid or exit the icing area, abort the mission, or initiate emergency landing.

The basic requirements for UAV ice sensors are that they must be low cost, small, lightweight, and efficient. Without the availability of a pilot to make a final assessment, unmanned aircraft must autonomously determine when the vehicle is entering and leaving icing conditions. Consequently, a high degree of automation is required [39], and the systems should operate as a primary automatic ice detection system. This underlines the importance of rapid response time, high sensitivity, and high accuracy of a UAV ice detection system. The absence of these characteristics can lead to hazardous false negatives (icing not detected) and power-wasting false positives (icing indicated when not present). A delayed activation of an IPS can lead to hazardous accumulations of ice. In addition to identifying icing conditions, it can be beneficial if ice sensors give an indication of icing rate. This information can be used to optimize IPS operation strategy to reduce power consumption.

Today, a wide range of ice detection concepts exist that can be grouped depending on the concept of their physical measurements [44]. The main concepts are detectors that identify existing ice accretions on the surface, atmospheric icing conditions, or performance degradation due to icing. The most common ice sensors identify ice accretions on a surface ("known icing"), with sensors typically located on the leading edge or on exposed probes. There are many possible physical effects that can be used for icing detection:

1. Optical systems observe changes in reflectivity between an iced and a clean surface. Onboard (gimbal) cameras directed at the wings may be used as well, but are less reliable.
2. Mechanical systems mostly rely on vibrating elements that change eigenfrequency during ice accumulation - and are typically used for manned aircraft and on large military UAVs [59]. Several detection methods involve microwaves, sonic, or ultrasonic waves that change their scatter or reflect differently in the presence of ice on a surface.
3. Electrical ice detection systems use changes in the properties of capacitance, impedance, or time-domain reflectometry of an iced surface.
4. Thermal detection systems measure changes in surface temperature due to the release of latent heat of the freezing droplets or by generating thermal signals and measuring the response [37].
5. Systems to identify atmospheric icing conditions detect the presence of supercooled droplets in the air. Such systems typically rely on optical array probes; light, detection, and ranging (LIDAR); radar; hotwire probes; or cloud spectrometers.
6. Last but not least, icing conditions can be identified by monitoring deviations from the flight performance [40]. The degradation of the aerodynamic performance affects the dynamics behavior of the aircraft which can be detected by suitable model-based estimation method and linked to potential icing conditions. Similarly, an increase of power consumption for propulsion may be linked to icing.

Note that systems may be limited by a minimum LWC threshold for detection. There are many more concepts of icing detection developed for manned aviation and for cold-climate wind energy applications with varying levels of maturity, and not all of them can be listed here [44]. The key takeaway is that while a multitude of physical methods can be used, the resulting sensors must be designed to meet the special requirements for UAVs.

Icing Environments

Icing environments describe what combinations of exposure time, LWC, and MVD can be typically expected in icing situations. This information is important for the design of aircraft and IPS. Four icing envelopes are used for the certification of passenger transport airplanes for icing conditions [25]. In the FAA Code of Federal Regulations 14 CFR Part 25 Appendix C, two in-cloud icing envelopes are described that indicate the probable maximum values of LWC (Figure 11). The continuous maximum icing envelope for stratiform clouds indicates that this type of icing can occur between sea level and altitudes of up to 22,000 feet with a horizontal extent of 17.4 nautical miles. The intermittent maximum icing envelope represents icing between altitudes of 4,000 and 22,000 feet in cumuliform clouds with a horizontal extent of 2.6 nautical miles. A third envelope called "take-off" has reduced LWC values from the surface to 1,500 feet above ground level. In recent years a fourth envelope, covering SLD icing, including freezing drizzle and freezing rain events, has been added as Appendix O. In addition, two alternate icing envelopes for helicopters that operate below 10,000 feet exist. The envelopes exhibit lower maximum LWC values that have been found by two studies on low-altitude icing environments [60, 61]. More research is needed to identify which envelopes are best suited for UAVs, depending on their operations. Note that these envelopes are averaged over a standardized distance, and conditions can occur outside of these envelopes.

FIGURE 11. **14 CFR Part 25 Appendix C, icing envelopes: continuous maximum (left) and intermittent maximum (right). Lines indicate a combination of MVD and LWC for a given air temperature.**

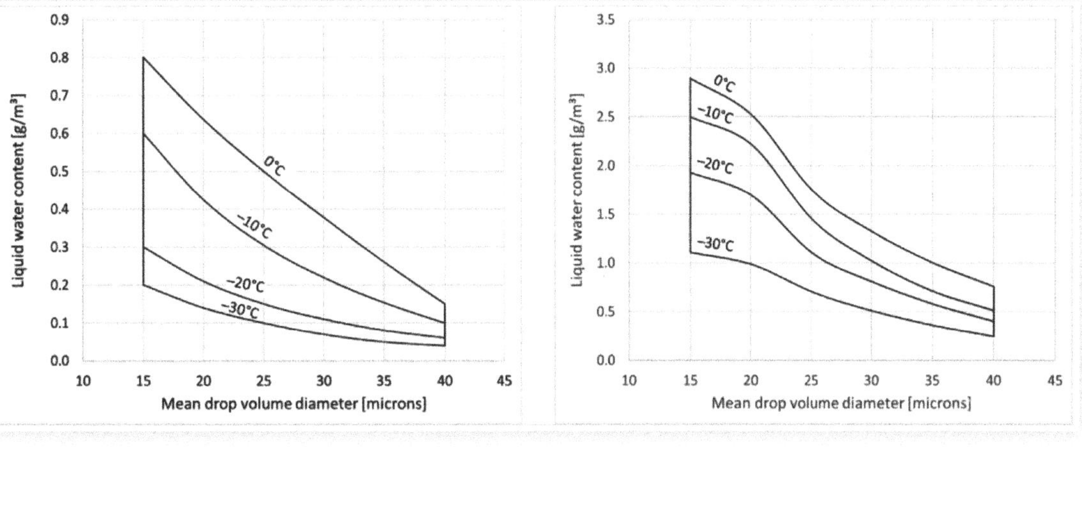

The icing environments encountered by UAVs can vary significantly depending on the UAV type. Large HALE UAVs may operate at altitudes similar, or even higher, than most manned aircraft, whereas small UAVs will operate typically in close proximity to the ground. The exposure to icing conditions may also be significantly longer for UAVs (e.g., surveillance missions). It is currently unclear how these special icing environments differ from the icing envelopes for manned aircraft. Of particular interest is icing near ground levels, since small UAVs are expected to operate in predominantly that regime. Experiences from wind turbine icing and electric power transmission lines show that the topography can significantly affect icing conditions [42]. Local variability can be significant due to the influence of terrain and open bodies of water. More research is required to study if this variability exceeds the certification envelopes.

Another aspect that needs investigation is how to assess the icing severity for small UAVs. As discussed above, icing conditions that are considered as light for a larger manned aircraft may be more severe for small UAVs. Similar discussions for manned aircraft of different sizes have been conducted in the past, and are still ongoing [47]. Similarly, smaller droplets that are considered to be less relevant for manned aircraft (i.e., left of the Appendix C, icing envelopes) may need to be considered for UAVs.

Icing Nowcasting and Forecasting

Nowcasting refers to a short-term weather forecast that is typically based on observations. Forecasting is based on numerical weather-prediction models providing long-term estimates. The knowledge of current and future icing conditions and other meteorological parameters (e.g., wind, temperature, precipitation, turbulence, and cloud cover) are important information for UAV operations and mission planning. In particular, such information is needed for optimized path-planning tools that find the most energy-efficient path of an aircraft to its destination [41]. A large variety of weather products exist that target the needs of manned aviation. Many of these products prove insufficient for UAV applications, especially for small UAV operations at low altitudes, due to the coarseness in resolution [62]. For example, the Current Icing Product and Forecast Icing Product provided by the United States National Weather Service estimate icing severities with a horizontal grid point spacing of 13 kilometers and a vertical spacing of 500 feet (approximately 150 meters) [63]. Another issue is that their icing-severity estimates are calibrated for manned aviation. The lack of icing nowcasting and forecasting data specifically designed for UAVs is likely to lead to conservative mission planning, which consequently limits the operational boundaries of UAVs [27]. This can significantly decrease the availability and overall value of a UAV application. Therefore, high-quality, UAV-tailored icing nowcasts and forecasts are required, which can provide sufficient accuracy and spatial and temporal resolution. For example, such models have been developed for wind turbines, providing high-resolution nowcasting and forecasting of icing in close proximity to the ground across Scandinavia [64].

IPS for UAVs

To mitigate the adverse effects of ice, various types of IPSs have been developed over the years. In manned aviation, three types of IPS are typically used: thermal systems, chemical systems, and mechanical systems. There are also other IPS concepts, which are less used, mostly due to their novel nature. This section discusses the most relevant IPS technologies along with their advantages and disadvantages.

In general, two different strategies exist for IPS: anti-icing and de-icing. The goal of anti-icing is to continuously protect a surface (e.g., wing or propeller) to avoid any ice accretions at all. The advantage of anti-icing is that there are no aerodynamic performance effects related to ice, but on the downside, this approach typically requires high resources. De-icing systems operate periodically and allow for a defined ice amount to build up (intercycle ice), which is then cyclically removed. Typically, de-icing systems require significantly less resources compared to anti-icing. However, there can be substantial aerodynamic performance penalties related to the intercycle ice. Aerodynamic forces contribute to ice shedding, but are dependent on the flight velocity [50]. Ice pieces that are shed by the de-icing system can damage downstream aircraft components (e.g., antennas, control surfaces) [53]. Dislodged ice that hits rotors/propellers or gets ingested by engines can cause substantial damage. The risk of ice shedding therefore needs to be considered for de-icing system designs.

Suitable IPS technologies for UAVs need to be lightweight, energy-efficient, and capable of autonomous operation. Furthermore, cost is a significant aspect and UAV IPS technologies need to be developed at a price that is acceptable to the manufacturers. The latter may be a challenge, in particular for smaller and cheaper UAVs.

Thermal

Thermal systems mitigate icing by supplying heat to critical airframe surfaces (Figure 12). The most common thermal systems are electrothermal systems and hot-air systems. Electrothermal systems are based on electric resistors that warm up when a voltage is applied. Hot-air systems utilize hot bleed air from jet engines that is distributed with piccolo tubes to the leading edge of the aircraft. Hot-air systems are mature, can be very efficient, but come with the risk of runback icing and an engine performance penalty [53].

Electrothermal systems are a mature technology, which are lightweight and simple. Heaters are made of carbon composites, conductive carbon nanotube coatings, or metal resistors. Electrothermal IPS exist for UAV

FIGURE 12. **Operation modes of a thermal IPS. Anti-icing provides continuous heat to the surfaces and prevents any ice formation, whereas de-icing systems run periodically and allow for ice accretions in between cycles.**

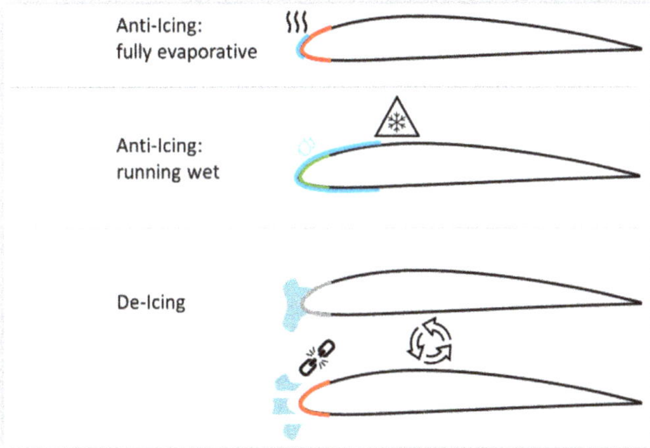

lifting-surfaces [38], rotors [54], and pitot tubes. Such systems may also be used to develop retrofittable IPS that can be adapted to UAVs without replacing existing hardware such as wings. Electrothermal systems are quick, robust, and lightweight but have a relatively high energy consumption with high power requirements for heating [53].

Anti-Icing Thermal anti-icing systems continuously heat the protected surfaces to prevent any ice accretions. Such systems can operate in two ways, which are characterized by the surface temperatures (Figure 12). Fully evaporative systems raise the surface temperature to a point that all impinging liquid water evaporates completely within the heated area. Running-wet systems prevent the incoming droplets only from freezing. Consequently, a liquid-water film develops on the surface that is flowing downstream as so-called runback water. Most running-wet systems only heat a limited area near the leading edge to reduce power consumption, which introduces the risk of runback icing [53]. When the water film flows aft of the heated zone, it can refreeze as runback ice and may form spanwise ice ridges that are associated with high aerodynamic penalties - potentially even worse than the penalties that would arise from an unprotected surface [9]. Higher surface temperatures of the heated zone accelerate evaporation rates, which reduces the risk for runback icing.

Figure 13 shows an example for the required heat loads of an anti-icing system in fully evaporative and running-wet mode on a typical UAV airfoil [65]. For full evaporation, the area to be heated is limited to the impingement zone; whereas for running wet the area is limited by the point where the water film evaporates. In practice, it is difficult to provide heat exclusively to the impingement zone or the water film zone, as the zone limits shift with different meteorological and flight conditions. IPS designs need to assess the range

FIGURE 13. **Example for heat requirements along the chord of an airfoil for fully evaporative and running-wet anti-icing.**

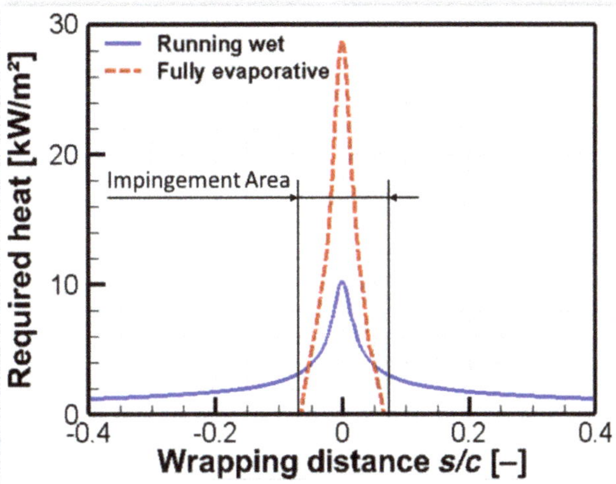

of expected icing conditions and design the system in such a way that the worst-case scenarios are covered.

The amount of required power for anti-icing strongly depends on the air temperature. Running-wet systems typically require less energy at temperatures close to freezing than evaporative systems. This is because the temperature difference they need to provide is comparatively small. At very low temperatures however, fully evaporative systems may require less energy as the heating area is much smaller [65].

The choice between a running-wet or a fully evaporative anti-icing system cannot be solely decided by power efficiency considerations. One critical aspect is the surface temperatures that are required for full evaporation of the droplets and their influence on the airframe. UAV structures can be built from a wide range of materials, including foams, glass fibers, carbon fibers, or metals. High anti-icing temperatures may lead to the loss of integrity of the materials or resins used in composites.

De-Icing Thermal de-icing systems operate periodically and allow for a defined ice amount to build up (intercycle ice), which is then removed during a short heating cycle (Figure 12). The elimination of the intercycle ice accretions is achieved by two main processes: melting and shedding [53]. During the heating cycle, a meltwater film forms at the interface between the heated surface and the ice layer. This water film decreases the adhesive forces between the ice and the surface. The overlaying ice sheds in the presence of sufficiently large aerodynamic forces and is therefore dependent on the flight velocity. A de-icing system is energy efficient when it minimizes melting and maximizes shedding, and typically requires less energy than anti-icing. Some de-icing designs utilize a continuously heated parting strip at the leading edge. This results in two separated ice accretions on the upper and lower side of the airfoil. During de-icing, ice sheds individually and more efficiently, since the aerodynamic drag forces are more efficiently contributing to overcome the surface adhesion of the ice.

For rotary-wing UAVs, anti-icing is often preferred to avoid the aerodynamic penalties. Static lifting surfaces have higher tolerances to ice penalties, which allows for the use of de-icing systems that require less energy compared to anti-icing. However, runback icing is still a risk for de-icing systems, because of the water film that is generated during the melting processes. Designing de-icing systems is a more complex process compared to anti-icing since it includes a larger number of parameters (e.g., intercycle time, heating time, and heating load) that need to be optimized for different meteorological icing conditions. There are intricate interactions between these parameters and total power consumption of the vehicle. For example, the intercycle ice accretions increase drag and decrease lift - which is compensated by increased thrust. On the one hand, the longer the intercycle duration is, the larger the penalties are and the higher the fuel consumption will be. On the other hand, shorter cycles lead to increased de-icing loads and increased runback icing risk due to higher meltwater production. Also, if the de-icing cycles are not timed well, the intercycle ice may not shed properly and keep building up. Balancing these aspects is a difficult task and requires thorough design and knowledge of the icing severity in order to avoid putting the aircraft in danger.

Mechanical

Mechanical IPS dislodges ice accretions on the surface and consequently breaks the adhesive forces between the ice and the surface. These systems work as de-icers and typically require a minimum threshold amount of ice to build up before they can be effectively used.

Pneumatic Boots The most common mechanical IPS is pneumatic de-icing boots, which are made of rubberlike material bonded to the airframe that can be inflated. The expansion of the boots dislodges the ice and aerodynamic forces - or centrifugal forces in case of a rotor - remove it from the surface. De-icing boots are commonly used on wings and propellers of light manned aircraft. Typically, auxiliary pneumatic systems are required to generate the required overpressure and vacuum for inflation and deflation of the boots.

Pneumatic boots are a mature technology that is simple, lightweight, energy efficient, and easily retrofittable [53]. However, the system can be responsible for significant aerodynamic performance degradation. The boots disrupt the airfoil geometry and generate additional drag, even in clean conditions [66]. Furthermore, intercycle ice and residual ice can significantly affect the aerodynamic performance [67]. The de-icing efficiency is decreased at low speeds, in icing conditions with large droplets with runback icing, and for surface roughness [68]. The lifetime of de-icing boots is typically limited due to erosion, abrasion, and weathering (by ozone and ultraviolet light).

Electromechanical A newer de-icing technology is electromechanical systems. These systems use electromechanical actuators to accelerate the outer skin of the airframe to de-ice the surface. The accelerations occur with a high frequency that displaces the skin with very low amplitudes. The resulting impulses generate a shock wave inside the structure that breaks the ice adhesion forces. There are several key technologies and patents of electromechanical IPS that differ mostly in how the acceleration of the surface is achieved: electromechanical expulsion de-icing systems, electric pulse ice protection system, electroimpulsive de-icing, sonic pulse electroexpulsive de-icer, and electroexpulsive separation system [69]. All of these technologies have low average power consumption and are lightweight; however, some systems have high instantaneous power requirements, may increase structural fatigue, may be prone to erosion, have the same intercycle and residual ice penalties as pneumatic boots, and are mostly not as mature as other technologies.

Chemical

The principle of all chemical IPS is to depress the freezing point of water sufficiently to prevent ice formation. Such chemicals are commonly used for de-icing of aircraft on the ground, prior to takeoff. Typically, a heated glycol solution is sprayed on the aircraft to remove ice, snow, and frost. Special fluids, engineered to stick to the surfaces, can provide temporary anti-icing protection during taxiing and the initial stages of takeoff, but have environmental risks [70].

Chemical systems are also used for in-flight ice protection, mostly in general aviation and often referred to as weeping wings. These systems disperse a freezing point depressant (FPD) fluid across the airframe through small orifices on the leading edge, which covers and protects the entire wing. The system can provide both de-icing and anti-icing capabilities and is commonly used for the protection of aircraft wings, propellers, and engine inlets of normal category airplanes. FPD systems consist of a small electrical pump and a fluid storage tank. The endurance of the system is limited by the amount of liquid carried. The system is required to be exercised regularly to ensure proper distribution of the fluid through the porous panels. If the system is not exercised as required, additional maintenance procedures may be required prior to flying in icing conditions to ensure expected performance. FPD systems require very low power and can operate without residual or runback icing [53]. The technology is mature with a long lifetime and has been implemented on UAVs (e.g., in some versions of the Predator) [17].

Icephobicity

Icephobicity, analogous to hydrophobicity, is a relatively new term that is used to describe a material property that is resistant to icing. The term is not well defined but generally includes three properties: low adhesion between ice and the surface, prevention of ice formation, and a repellent effect to supercooled droplets [71]. Icephobicity requires special material properties and is - as often misconceived - not correlated to superhydrophobicity [72]. Icephobic coatings can be applied to surfaces by two different methods: via coatings as either painted, sprayed, dipped, or brushed onto a surface as a retrofit or via modifying the surfaces themselves using vacuum deposition methods, physical etching techniques, etc. [73].

Icephobic coatings may be considered as the holy grail of IPS. They are passive systems, not requiring any form of power during flight, and are very lightweight. Coatings can also be paired with active systems to increase their efficiency [73]. For this reason, considerable research efforts are ongoing especially in the fields of aircraft icing and wind turbine icing. While there are promising results, the technology has yet to be proven outside of laboratory conditions. Several challenges exist that are related to real-world conditions [73]. Small droplet sizes combined with high velocities, for example, can lead to a penetration of the coating layers resulting in increased ice adhesion forces. However, the largest concern is related to the resistance to erosion by ice crystals, raindrops, sand, and insects. This can lead to a rapid degradation of the icephobic properties.

Further research and development are required before icephobic coatings are ready to be implemented. Icephobicity is a promising technology for UAVs for the abovementioned reasons, and UAVs may be well suited as test beds for the new technology. UAVs require smaller areas to be coated, allow frequent maintenance intervals for reapplication, and can be flight tested without endangering human lives. Particularly, rotors and propellers could benefit from a reduction of adhesive forces, leading to an increase in ice-shedding efficiency in combination with centrifugal forces.

Novel Concepts

There is ongoing research to develop new and superior IPS technologies. There are many physical concepts that may be exploited for icing protection, but that have not yet been proven on a wider scale. Three notable examples that are relevant for UAVs shall be introduced here briefly. Shape memory alloys change their shape when heated, similar to conventional thermal expansion [74]. The deformation of shape memory alloy is enhanced to a level where it is enough to mechanically break the ice. Ultrasonic waves induced in the aircraft skin via piezoelectric actuators can generate forces that are strong enough to break ice accretions [75]. Microwaves can be used on special coatings to generate heat for anti-icing and de-icing solutions mainly for wind energy and turboprops [76]. Hybrid systems are systems that combine passive coatings and active systems [73]. All of these systems are lightweight and efficient - making them well suited for UAV applications. However, further research is required before these concepts are ready for implementation on a larger scale.

Method and Tools

The design process of an all-weather capable UAV suited for icing conditions is analogous to manned aircraft. As such, there are several mature methods and tools available to demonstrate that the aircraft can operate safely and reliably in icing conditions. In the beginning of the design process, numerical simulation tools are a fast and cheap method to investigate different designs and to identify worst-case scenarios. Due to the inherent uncertainties of computational simulations, experimental tests are required at later stages in the design. Specialized wind tunnels have been developed that can replicate in-cloud icing environments under laboratory settings, which can verify ice shapes and IPS performances. Flight tests, with artificial ice shapes or in natural icing conditions, are the ultimate test to ensure the icing handling qualities of an aircraft.

Numerical Simulation

Numerical simulation methods for aerodynamics have been developed since the 1970s and 1980s and are a key tool for the development of any aircraft - manned or unmanned. They contribute to the design and certification process from an early stage on to investigate different icing conditions and IPS designs [14]. Simulations can investigate many cases and designs in short timeframes and at lower costs than experimental methods. As such, a large selection of simulation models has been developed, aiming to capture all relevant icing processes such as ice accretion on static and rotating surfaces, engine ice-crystal ingestion, ice shedding, SLDs, anti-icing, de-icing, etc.

Most icing codes use an iterative approach to simulate the icing process in four steps (Figure 14) [77]. First, the flow field around the aircraft or airfoil is simulated. Today, most codes use three-dimensional (3D) computational fluid dynamics (CFD) methods that solve the Navier-Stokes equations to derive the flow field. Some older codes exist and are still in application. These codes are of lower order and use panel methods instead. Second, the droplet impingement on the surface is simulated, resulting in information about the amount of water caught. Droplet trajectories are calculated based on the flow field results with either a Lagrangian or Eulerian approach. Third, the energy and mass balance is solved on the surface to calculate the amount of liquid water turning into ice. This step is straightforward for rime ice, where all droplets are considered to freeze on impact, but is a complex task when a water film is present on the surface (e.g., during glaze icing or IPS operation). Fourth, the new surface geometry is calculated, based on the ice growth rates. Tools that are based on CFD methods need to either generate a new computational mesh or adapt the existing mesh to the ice growth.

As with all numerical methods, validation is a key requirement to ensure the simulation results match with real-world data. For manned aviation, a large number of experimental ice shapes for wings and airfoils are available for validation (e.g., [78]). Substantially, less data is available for rotors and propellers. However, most of these validation data are not applicable to UAVs, as they are conducted at higher velocities with larger wing sizes. Very little data is currently available that is suitable for the validation of UAVs that are operating at low Reynolds numbers [31, 79]. Consequently, there is a significant uncertainty associated with using icing tools developed for manned aviation for UAV applications. One of the main limitations is related to the low Reynolds flow regime of most - especially smaller - UAVs. CFD codes are struggling with the prediction flow features such as laminar-turbulent transition, laminar separation bubbles, or separation, even in clean cases [80]. The modelling of these in the presence of rough iced surfaces adds an additional layer of complexity. It is currently unknown how the existing icing models and codes perform for low Reynolds numbers, due to the lack of benchmark studies and validation data.

In the literature, three icing codes have been used on UAVs so far. LEWICE2D - a panel-method-based icing code - has been developed by the National Aeronautics and Space Administration (NASA) [78]. Preliminary results indicate that LEWICE2D may be suited for the prediction of anti-icing IPS loads and rime ice shapes, but has shown deviations from glaze ice experiments [33, 79]. A CFD implementation of LEWICE that is capable of 3D simulations exists and may be better suited for the complex icing cases, but has not been applied to UAVs yet. FENSAP-ICE is a modern CFD icing code that is being commercially developed by ANSYS [81]. The little experimental data, which is available, indicates that the code may perform well for UAVs, also for glaze conditions [38, 79]. A novel, morphogenetic modelling approach is been developed at the National Research Council Canada [82]. The code has been used to investigate Reynolds number effects and

FIGURE 14. Simulation of ice accretion in four steps: (1) airflow, (2) droplets, (3) ice growth, (4) remeshing.

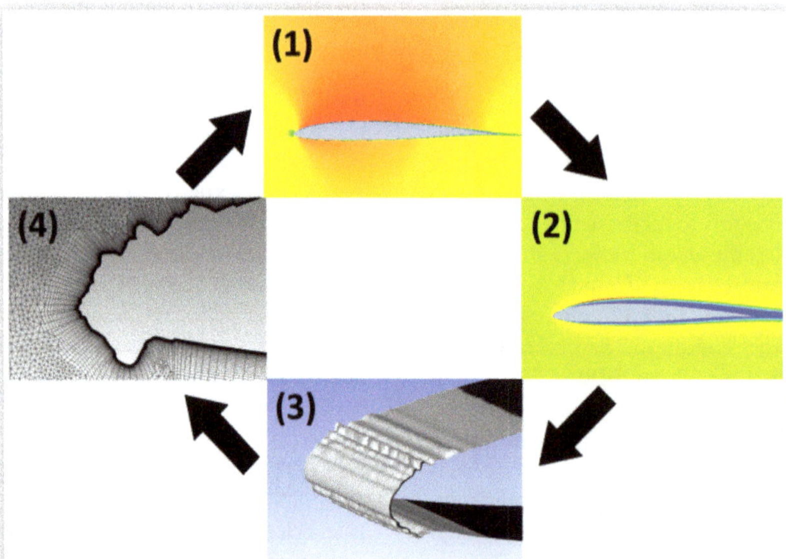

performance degradation on UAV airfoils, but is also lacking validation for that flow regime [32, 46].

Icing Wind-Tunnel Experiments

Experimental icing wind-tunnel (IWT) tests are a key element in the design of any aircraft with all-weather capabilities. An IWT allows for the investigation of icing under clearly defined laboratory conditions, and can be seen as an intermediate step between numerical simulation and flight tests [14]. Typically, two types of tests are conducted. First, IWTs are used to verify the worst-case ice shapes predicted by numerical tools. The resulting ice geometries are then documented and artificially replicated, for example, by 3D printing. These artificial ice shapes are subsequently used to investigate the aerodynamics performance penalties in conventional wind tunnel tests or in-flight tests. Second, IPS and ice detection systems are tested in IWTs to validate their functionality and verify their energy requirements.

IWTs are built similarly to conventional wind tunnels and can be designed as open-circuit or closed-circuit tunnels. The two main additional components are a spray system to inject liquid droplets and a facility to refrigerate the flow. Closed-circuit and climate-controlled tunnels cool the flow that is circulating in the tunnel, whereas open-circuit tunnels are in environments (e.g., climate chambers) with freezing temperatures. Special facilities exist for icing on rotorcraft [21]. A large variety of IWT designs exist, with test sections ranging from several meters to decimeters and wind speeds mostly below Mach 0.4.

For conventional (non-icing) wind-tunnel testing, downscaled models are commonly used in manned aviation due to size restrictions of the tunnels. In this non-icing case, only two nondimensional parameters (Reynolds and Mach numbers) need to be matched to obtain similarity between the scaled and the full-scale model. With the introduction of droplets into the flow, the number of nondimensional parameters increases excessively to about 18, which is impossible to meet in practical terms [14]. In current practice for aircraft icing, either scaled models are used or limited segments of the airframe are tested at full scale [83].

In contrast, UAVs can often be tested at 1:1 scale, since they are smaller in size and operate at lower airspeeds. This is a large advantage since IWT test results are directly representative of real-world conditions. In addition, their smaller size can allow UAVs to be more flexible in the selection of IWT facilities, opening testing for smaller tunnels which are mainly used for research or other fields of icing (e.g., wind turbine icing). An overview of international IWT facilities can be found in [43].

Flight Tests

The most accurate tool to test the ability of an aircraft to survive in icing conditions is to fly in natural or simulated icing conditions. This process is - for obvious reasons - very expensive and dangerous and is therefore conducted in the final stages of an icing certification process. One practical difficulty is to find icing clouds within the airspace reserved for testing. Aircraft icing certification campaigns may be significantly delayed if the right weather conditions are not available. Icing flight campaigns typically include tests on the aircraft's control, trim characteristics, stability, and stall [14]. Tests can be performed with artificial ice shapes or in natural icing conditions. Spray tanker aircrafts, which generate an artificial icing cloud by spraying water from nozzles behind the aircraft (e.g. [84]), may be used if in-flight collision safety concerns can be resolved.

Flight tests have a much lower risk threshold to be conducted with UAVs compared to manned aircraft, since no humans are directly involved. In the future, natural in-flight icing tests may develop into an affordable alternative to IWT tests in areas where natural icing conditions occur regularly [6, 7].

Regulations

UAVs are a relatively recent addition to the aviation system and their integration is an ongoing process. There are many obstacles and challenges when it comes to safely integrating civil UAVs into a nonsegregated airspace and to develop a robust regulatory framework [85]. Icing is only one of many concerns and may not rank as high in urgency as other questions (e.g., regarding integration into the air traffic management system or enforcing no-fly zones). Nonetheless, icing remains a relevant question, especially since icing on UAVs can be a public safety hazard when operations are conducted in populated areas. Aviation safety agencies like FAA, Transport Canada Civil Aviation, and European Aviation Safety Agency have recognized the importance of UAV icing hazards and the need for regulation. However, to this date, only Canada has introduced explicit rules regarding UAV operations and icing conditions in their new regulations from early 2019 [86]. The new UAV operation regulations in Europe, valid from 2020 onwards, do not yet include any notes on icing [87].

Civil UAV icing operations and certification will most likely be eventually addressed with the same rigor as civilian manned aircraft [62]. For manned civil aviation, icing regulations for certification are stated in 14 CFR Part 23/25/29/33 for normal airplanes/transport airplanes/rotorcraft/aircraft engines (and CS-23/23/29 in Europe) [25]. The regulations define specific icing conditions and flight scenarios that must be met in order for the aircraft to be approved for icing. It seems fathomable that comparable certification rules will be introduced for UAVs above a certain weight limit (e.g., 25 kilograms/55 pounds) in the foreseeable future. Rules may be also based on third-party risk and operational areas. The regulations will need to reflect the differences to manned aviation and the UAV-specific topics discussed at length in the scope of this work. Different regulations may be developed for UAVs with and without IPS. To this date no uniform regulations exist, and more research is needed on the topic before UAV icing can be addressed appropriately. The introduction of a certification process is most likely going to significantly increase costs and workload for the development of all-weather capable UAVs.

For military UAV systems, the North Atlantic Treaty Organization (NATO) Standardized Agreement STANAG 4671 gives airworthiness requirements for the operation of UAVs in NATO airspace [88]. The document, first released in 2009, addresses several topics related to icing on UAVs and IPS. In particular, the document requires demonstration that all-weather capable UAVs are qualified to operate in continuous maximum and intermittent maximum icing conditions.

Summary

Icing on UAVs imposes a key limitation to their operational envelopes and reduces their value for commercial, civil, and military applications. Today, there is a lack of tools and technologies to deal with sthe icing hazards on UAVs. There are no specialized tools for icing nowcasting and forecasting, no mature ice-detection systems, and no well-proven IPS for the most common UAV sizes. This situation has risen from the circumstance that the UAV is an emerging technology - especially for commercial applications - and that cold-climate capabilities have been a low priority until now for the military. The cost for development of suitable icing technologies compared to unit costs is also an obstacle. Substantial research and engineering efforts are required to unlock the full potential of many UAV applications discussed today. New technology developments as well as transfer of existing technologies from manned aviation are necessary to ensure that UAVs can operate economically, safely, reliably, and repeatedly in icing conditions in the future.

SAE EDGE™ Research Reports

SAE EDGE™ Research Reports, like the present report on "Unsettled Topics in Unmanned Aerial Vehicle Icing," are intended to push further out into still unsettled areas of technology of interest to the mobility industry. SAE launches these reports before attempting to form a joint working group, let alone a cooperative research program or a standards committee.

SAE EDGE™ reports are intended to be quick, concise overviews of major unsettled areas where vital new technologies are emerging. An unsettled area is characterized more by confusion and controversy than established order. Early practitioners must confront an absence of agreement. Their challenge is often not to seize the high ground but to find common ground. These scouting reports from the frontiers of investigation are intended merely to begin the process of sorting through critical issues, contributing to a better understanding of key problems, and providing helpful suggestions about possible next steps and avenues of investigation.

SAE EDGE™ Research Reports, therefore, are fundamentally distinct from the more formal working groups approach and far removed from the more mature research program and standard's development process.

Next Steps for Unsettled Topics in Unmanned Aerial Vehicle Icing

This publication should be considered only as a first step toward clarifying the issues around UAV icing. The intention behind this and other SAE EDGE™ Research Reports is to start a dialogue among interested parties on important industry-wide topics that require further attention. The expectation is that these explorations of unsettled areas of technology will lead to the formation of working groups and, ultimately, committees that can address and resolve the issues they raise, producing a framework for developing a common vocabulary of definitions, best practices, protocols, and standards needed to support continued progress toward safer and more innovative products.

The experts' collaboration that gave rise to this publication demonstrated a great willingness on the part of the industry to define the terminology, procedures, and eventually the standards needed to address UAV icing to move ahead as quickly and efficiently as possible.

Recommendations

This SAE EDGE™ Research Report on "Unsettled Topics in Unmanned Aerial Vehicle Icing" identifies the following key topics for further pursuit, both through continued informal discussions among industry practitioners and through more formal working groups:

Short term

1. Development of robust ice-detection systems for UAVs that are low cost, lightweight, low energy-consumption, and sensitive enough for light icing conditions.
2. Transfer of mature IPS technologies (mitigation systems, pitot tubes, etc.) from manned aviation to UAVs to extend the operational envelope of unmanned aircraft to icing conditions.
3. Identification of existing icing nowcasting and forecasting products that could be used for risk management and planning of UAV operations in proximity to the ground and at superhigh altitudes.
4. Validation of the numerical icing tools developed for manned aviation to UAV flying conditions. The availability of reliable simulation tools is a cost- and time-saving element in the design of UAV IPS.

Long term

1. Development of icing detectors that can measure ice-accretion rates and identify SLD conditions.
2. Maturation of novel IPS technologies for UAVs with the aim to reduce weight, power, and cost; this will allow to all-weather capabilities for a wider range of UAV applications.
3. Development and improvement of the accuracy of icing nowcasting and forecasting products aimed for risk management and planning of UAV flights in proximity to the ground and at superhigh altitudes.
4. Performance of flight campaigns to establish icing envelopes for UAVs flying at low altitudes.
5. Design of UAV autopilots that can detect and adapt to icing conditions.
6. Development of improved icephobic materials and integration methods for UAVs.
7. Implementation of internationally standardized regulations for UAVs in icing conditions; this includes considering the creation of a specialized icing envelope for UAVs, similar to the existing regulations for manned aircraft.

Acronyms

3D - Three-dimensional
BVLOS - Beyond visual line of sight
CFD - Computational fluid dynamics
CFR - Code of Federal Regulations (FAA)
FAA - Federal Aviation Administration
FPD - Freezing point depressant
HALE - High-altitude long-endurance
IPS - Ice protection system
IWT - Icing wind tunnel
LWC - Liquid water content
MVD - Median (drop) volume diameter
NASA - National Aeronautics and Space Administration
NATO - North Atlantic Treaty Organization
RPAS - Remotely piloted aircraft systems
SLD - Supercooled large droplets
UAV - Unmanned aerial vehicle
VLOS - Visual line of sight

Acknowledgments

Recognition should go first to all the participants, many of whom also provided feedback on the draft version of this publication. Without their input and initiative, this SAE EDGE™ Research Report would not have been possible.

Ali Benmeddour, National Research Council Canada
Ben Bernstein, Leading Edge Atmospherics
Martin Fengler, Meteomatics
David A. Funkhouser, Boeing Insitu
Morten Hansbø, Norwegian Defence Research Establishment
Torbjørn Houge, Maritime Robotics
Philip McGillivary, United States Coast Guard
Riku Niemenmaa, Finnish Defence Forces
Isik Ozcer, ANSYS
Paul Pellicano, Federal Aviation Administration
Dennis Regnier, CAV Systems
Kim Lynge Sørensen, UBIQ Aerospace
Rune Storvold, NORCE Norwegian Research Centre
Kevin Yugulis, Battelle Memorial Institute

The authors of this document together with the SAE Team responsible for its creation join in expressing our deepest appreciation to all the individuals mentioned above.

Richard Hann and Tor Arne Johansen, Norwegian University of Science and Technology (NTNU)

In addition, we would like to acknowledge the research funded on UAV icing by the Research Council of Norway via the Centre for Integrated Remote Sensing and Forecasting for Arctic Operations (CIRFA) and the Centre for Autonomous Marine Operations and Systems (NTNU-AMOS).

References

1. Shakhatreh, H., Sawalmeh, A.H., Al-Fuqaha, A., Dou, Z. et al., "Unmanned Aerial Vehicles (UAVs): A Survey on Civil Applications and Key Research Challenges," *IEEE Access* 7:48572-48634, 2019.
2. IPCC, "IPCC Special Report on the Ocean and Cryosphere in a Changing Climate," 2019, https://www.ipcc.ch/srocc/cite-report/.
3. MarketsandMarkets, "Military Drones Market by Type, Application (ISRT, Delivery and Transportation, Combat Operations, Battle Damage Management), Range (VLOS, EVLOS, BLOS), Propulsion Type, Launching Mode, Endurance, MTOW, Region - Global Forecast to 2025," Market Report, - Retrieved November 2019, www.marketsandmarkets.com/Market-Reports/military-drone-market-221577711.html.
4. QY Research, "Global Commercial Drones Market Report, History and Forecast 2014-2025, Breakdown Data by Manufacturers, Key Regions, Types and Application," Market Report, Retrieved November 2019, www.qyresearch.com/index/detail/1140178/global-commercial-drones-market.
5. Federal Aviation Administration, "FAA National Forecast 2019-2039," Report, USA, 2019.
6. Bernstein, B.C., Wolff, C.A., and McDonough, F., "An Inferred Climatology of Icing Conditions Aloft, Including Supercooled Large Drops. Part I," *Journal of Applied Meteorology and Climatology* 46:1857-1878, Jan. 2007.

7. Bernstein, B.C. and Le Bot, C., "An Inferred Climatology of Icing Conditions Aloft, Including Supercooled Large Drops. Part II: Europe, Asia, and the Globe," *Journal of Applied Meteorology and Climatology* 48:1503-1526, 2009.
8. Gent, N.P.D. and Cansdale, J.T., "Aircraft Icing," *Philosophical Transactions of the Royal Society of London A: Mathematical, Physical and Engineering Sciences* 358:2873-2911, 2000.
9. Bragg, M.B., Broeren, A.P., and Blumenthal, L.A., "Iced-Airfoil Aerodynamics," *Progress in Aerospace Sciences* 41:323-362, 2005.
10. Janjua, Z.A., Turnbull, B., Hibberd, S., and Choi, K.S., "Mixed Ice Accretion on Aircraft Wings," *Physics of Fluids* 30, 2018.
11. Bernstein, B.C., Ratvasky, T.P., Miller, D.R., and McDonough, F., "Freezing Rain as an In-Flight Icing Hazard," NASA/TM-2000-210058, 2000.
12. Mason, J.G., Strapp, J.W., and Chow, P., "The Ice Particle Threat to Engines in Flight," in *44th AIAA Aerospace Sciences Meeting*, Reno, 2006, 2445-2465.
13. Farzaneh, M., *Atmospheric Icing of Power Networks* (Dordrecht, Netherlands: Springer, 2008).
14. Kind, R.J., Potapczuk, M.G., Feo, A., Golia, C., and Shah, A.D., "Experimental and Computational Simulation of In-Flight Icing Phenomena," *Progress in Aerospace Sciences* 34:257-345, 1998.
15. Lynch, F.T. and Khodadoust, A., "Effects of Ice Accretions on Aircraft Aerodynamics," *Progress in Aerospace Sciences* 37:669-767, 2001.
16. Fajt, N., Hann, R., and Lutz, T., "The Influence of Meteorological Conditions on the Icing Performance Penalties on a UAV Airfoil," in *8th European Conference for Aeronautics and Space Sciences (EUCASS)*, Madrid, 2019.
17. Leary, W.M., "We Freeze to Please: A History of NASA's Icing Research Tunnel and the Quest for Flight Safety," National Aeronautics and Space Administration (NASA), USA, 2002.
18. Green, S.D., "A study of U.S. Inflight Icing Accidents and Incidents, 1978 to 2002," in *44th AIAA Aerospace Sciences Meeting*, Reno, 2006.
19. Neel, C.J. and Bright, L., "The Effect of Ice Formation on Propeller Performance," NACA Technical Note NACA/TN/2212, 1950.
20. Flemming, R.J., "The Past Twenty Years of Icing Research and Development at Sikorsky Aircraft," in *40th AIAA Aerospace Sciences Meeting and Exhibit*, Reno, 2002.
21. Cao, Y. and Chen, K., "Helicopter Icing," *Aeronautical Journal* 114:83-90, 2010.
22. De Souza, J.R.B., Lisboa, K.M., Cerqueira, I.G., Zotin, J.L.Z. et al., "Conjugated Heat Transfer Analysis of Heated Aeronautical Pitot Probes with Flight Tests Experimental Validation," *Heat Transfer Engineering* 36:991-1000, 2015.
23. Gardner, L., Moon, G., and Whyte, R.B., "Aircraft Carburetor Icing Studies," SAE Technical Paper 710371, 1971, https://doi.org/10.4271/710371.
24. Pfeifer, G., "Aircraft Engine Icing, Technical Summary, Ice Test for Aircraft Engines," AGARD CP 236, 1978.
25. Federal Aviation Administration, "14 CFR Parts 25 and 29, Appendix C, Icing Design Envelopes," DOT/FAA/AR-00/30, 2002.
26. New Zealand Civil Aviation Authority, *Aircraft Icing Handbook* (Lower Hutt, New Zealand, 2000.)
27. Siquig, A., "Impact of Icing on Unmanned Aerial Vehicle (UAV) Operations," Naval Environmental Prediction Research Facility Report, Naval Environmental Prediction Research Facility, Monterey, CA, 1990.
28. Peck, L., Ryerson, C.C., and Martel, C.J., "Army Aircraft Icing Cold Regions Research and Engineering Laboratory," Cold Regions Research and Engineering Laboratory, Ft Belvoir, VA, 2002.
29. Haulman, D.L., "U.S. Unmanned Aerial Vehicles in Combat, 1991-2003," Air Force Historical Research Agency, Air Force Historical Research Agency, Montgomery, AL, 2003.
30. Defence Safety Authority, "Service Inquiry: Loss of Watchkeeper (WK042) Unmanned Air Vehicle over Cardigan Bay in West Wales," 2017.
31. Williams, N., Benmeddour, A., Brian, G., and Ol, M., "The Effect of Icing on Small Unmanned Aircraft Low Reynolds Number Airfoils," in *17th Australian International Aerospace Congress*, Melbourne, AIAC, 2017.
32. Szilder, K. and Yuan, W., "In-Flight Icing on Unmanned Aerial Vehicle and Its Aerodynamic Penalties," *Progress in Flight Physics* 9:173-188, 2017.
33. Hann, R., "UAV Icing: Comparison of LEWICE and FENSAP-ICE for Ice Accretion and Performance Degradation," in *2018 Atmospheric and Space Environments Conference*, 2018, 1-8.
34. Hann, R., Wenz, A., Gryte, K., and Johansen, T.A., "Impact of Atmospheric Icing on UAV Aerodynamic Performance," in *2017 Workshop on Research, Education and Development of Unmanned Aerial Systems, RED-UAS 2017*, Linköping, 2017, 66-71.
35. Liu, Y., Li, L., Chen, W., Tian, W., and Hu, H., "An Experimental Study on the Aerodynamic Performance Degradation of a UAS Propeller Model Induced by Ice Accretion Process," *Experimental Thermal and Fluid Science* 102:101-112, 2019.
36. Yan, S., Opazo, T.I., Langelaan, J.W., and Palacios, J.L., "Experimental Evaluation and Flight Simulation of

Coaxial-Rotor Vehicles in Icing Clouds," Journal of the American Helicopter Society, 2019.

37. Sørensen, K.L., "Autonomous Icing Protection Solution for Small Unmanned Aircraft," NTNU Ph.D. thesis, 2016.
38. Hann, R., Borup, K., Zolich, A., Sorensen, K. et al., "Experimental Investigations of an Icing Protection System for UAVs," SAE Technical Paper 2019-01-2038, 2019, https://doi.org/10.4271/2019-01-2038.
39. Rashid, T., Mughal, U.N., and Virk, M.S., "Atmospheric Icing Sensors for UAV's," in *4th IEEE International Conference on Cognitive Infocommunications, CogInfoCom 2013 - Proceedings*, 2013, 725-728.
40. Rotondo, D., Cristofaro, A., Johansen, T.A., Nejjari, F., and Puig, V., "Robust Fault and Icing Diagnosis in Unmanned Aerial Vehicles Using LPV Interval Observers," *International Journal of Robust and Nonlinear Control* 29:5456-5480, 2019.
41. Hovenburg, A.R., Andrade, F.A.A., Hann, R., Rodin, C.D., and Johansen, T.A., "Long Range Path Planning Using an Aircraft Performance Model for Battery Powered sUAS Equipped with Icing Protection System," in NTNU Ph.D. Thesis: Andrade, F.A.A.: "Real-Time and Offline Path Planning of Unmanned Aerial Vehicles for Maritime and Coastal Applications," 2019.
42. Battisti, L., *Wind Turbines in Cold Climates* (Cham, Heidelberg, New York, Dordrecht, London: Springer, 2015).
43. IEA Wind Task 19, "Available Technologies for Wind Energy in Cold Climates," Technical Report, 2018.
44. Homola, M.C., Nicklasson, P.J., and Sundsbø, P.A., "Ice Sensors for Wind Turbines," *Cold Regions Science and Technology* 46:125-131, 2006.
45. Homola, M.C., Wallenius, T., Makkonen, L., Nicklasson, P.J., and Sundsbø, P.A., "The Relationship between Chord Length and Rime Icing on Wind Turbines," *Wind Energy* 13:627-632, 2010.
46. Szilder, K. and McIlwain, S., "In-Flight Icing of UAVs - The Influence of Reynolds Number on the Ice Accretion Process," SAE Technical Paper 2011-01-2572, 2011, https://doi.org/10.4271/2011-01-2572.
47. Jeck, R.K., "A Workable, Aircraft-Specific Icing Severity Scheme," in *36th AIAA Aerospace Sciences Meeting and Exhibit*, Reno, 1998.
48. Office of the Secretary of Defense, "Unmanned Aircraft Systems Roadmap 2005-2030," Technical Report, 2005.
49. Jeck, R., "The Upper and Lower Temperature Limits for Aircraft Icing," Flight Safety Research Section AAR-470 William J. Hughes Technical Center, Federal Aviation Administration, USA, 2004.
50. Scavuzzo, R.J., Chu, M.L., and Ananthaswamy, V., "Influence of Aerodynamic Forces in Ice Shedding," *Journal of Aircraft* 31:526-530, 1994.
51. Li, L., Liu, Y., Zhang, Z., and Hu, H., "Effects of Thermal Conductivity of Airframe Substrate on the Dynamic Ice Accretion Process Pertinent to UAS Inflight Icing Phenomena," *International Journal of Heat and Mass Transfer* 131:1184-1195, 2019.
52. Pellicano, P., Dumont, C., Smith, T., and Riley, J., "Propeller Icing Tunnel Test on a Full-Scale Turboprop Engine," FAA Technical Report DOT/FAA/AR-06/60, 2010.
53. Thomas, S.K., Cassoni, R.P., and MacArthur, C.D., "Aircraft Anti-Icing and De-Icing Techniques and Modeling," *Journal of Aircraft* 33:841-854, 1996.
54. Fengler, M., "Study of Propeller Icing Hazard in Mini-UAV Aviation," Retrieved November 2019, www.meteomatics.com/wp-content/uploads/2019/02/SOPHIA Final Report-public.pdf.
55. Vukits, T.J., "Overview and Risk Assessment of Icing for Transport Category Aircraft and Components," in *40th AIAA Aerospace Sciences Meeting and Exhibit*, Reno, 2002.
56. Chao, H., Cao, Y., and Chen, Y., "Autopilots for Small Unmanned Aerial Vehicles: A Survey," *International Journal of Control, Automation and Systems* 8:36-44, 2010.
57. Winter, A., Hann, R., Wenz, A., and Johansen, T.A., "Stability of a Flying Wing UAV in Icing Conditions," in *8th European Conference for Aeronautics and Space Sciences (EUCASS)*, Madrid, 2019.
58. Caliskan, F. and Hajiyev, C., "A Review of In-Flight Detection and Identification of Aircraft Icing and Reconfigurable Control," *Progress in Aerospace Sciences* 60:12-34, 2013.
59. Jackson, D.G. and Goldberg, J.I., "Ice Detection Systems: A Historical Perspective," SAE Technical Paper 2007-01-3325, 2007, https://doi.org/10.4271/2007-01-3325.
60. Masters, C.O., "A New Characterization of Supercooled Clouds below 10,000 Feet AGL," FAA Technical Report DOT/FAA/CT-83/22, 1985.
61. Jeck, R.K., "A New Data Base of Supercooled Cloud Variables for Altitudes up to 10,000 Feet AGL and the Implications for Low Altitude Aircraft Icing," FAA Technical Report DOT/FAA/CT-83/21, 1983.
62. Campbell, S.E., Clark, D.A., and Evans, J.E., "Preliminary UAS Weather Information Gap Analysis for UAV Operations," 2017.
63. Bernstein, B.C., McDonough, F., Politovich, M.K., Brown, B.G. et al., "Current Icing Potential: Algorithm Description and Comparison with Aircraft Observations," *Journal of Applied Meteorology* 44:969-989, 2005.
64. Gregow, E., Bernstein, B., Wittmeyer, I., and Hirvonen, J., "LAPS-LOWICE: A Real-Time System for the

Assessment of Low-Level Icing Conditions and Their Effect on Wind Power," *Journal of Atmospheric and Oceanic Technology* 32:1447-1463, 2015.

65. Hann, R., "UAV Icing: Comparison of LEWICE and FENSAP-ICE for Anti-Icing Loads," in *AIAA Scitech 2019 Forum*, San Diego, CA, AIAA, 2019.
66. Albright, A.E., Kohlman, D.L., Schweikhard, W.G., and Evanich, P., "Evaluation of a Pneumatic Boot Deicing System on a General Aviation Wing Model, NASA Technical Report NASA/TM-82363, 1981.
67. Broeren, A.P., Bragg, M.B., and Addy, H.E., "Effect of Intercycle Ice Accretions on Airfoil Performance," *Journal of Aircraft* 41:6165-6174, 2004.
68. Pellicano, P. and Riley, J.T., "Residual and Inter-Cycle Ice for Lower-Speed Aircraft with Pneumatic Boots," in *Collection of Technical Papers - 45th AIAA Aerospace Sciences Meeting*, Reno, Vol. 19, 2007, 13149-13156.
69. Lawson, C.P., "Electrically Powered Ice Protection Systems for MALE UAVs - Requirements and Integration Challenges," in *ICAS-Secretariat - 25th Congress of the International Council of the Aeronautical Sciences 2006*, Vol. 6, 2006, 3565-3573.
70. Corsi, S.R., Geis, S.W., Loyo-Rosales, J.E., Rice, C.P. et al., "Characterization of Aircraft Deicer and Anti-Icer Components and Toxicity in Airport Snowbanks and Snowmelt Runoff," *Environmental Science & Technology* 40:3195-3202, May 2006.
71. Hejazi, V., Sobolev, K., and Nosonovsky, M., "From Superhydrophobicity to Icephobicity: Forces and Interaction Analysis," *Scientific Reports* 3(1):2045-2322, 2013.
72. Jung, S., Dorrestijn, M., Raps, D., Das, A. et al., "Are Superhydrophobic Surfaces Best for Icephobicity?" *Langmuir* 27:3059-3066, 2011.
73. Huang, X., Tepylo, N., Pommier-Budinger, V., Budinger, M. et al., "A Survey of Icephobic Coatings and Their Potential Use in a Hybrid Coating/Active Ice Protection System for Aerospace Applications," *Progress in Aerospace Sciences* 105:74-97, 2019.
74. Gerardi, J.J., Ingram, R.B., and Catarella, R.A., "A Shape Memory Alloy Based De-Icing System for Aircraft," in *33rd Aerospace Sciences Meeting and Exhibit*, Denver, CO, 1995, 1-8.
75. Wang, Z., "Recent Progress on Ultrasonic De-Icing Technique Used for Wind Power Generation, High-Voltage Transmission Line and Aircraft," *Energy and Buildings* 140:42-49, 2017.
76. Feher, L. and Thumm, M., "Design of Avionic Microwave De-/Anti-Icing Systems," in *Advances in Microwave and Radio Frequency Processing - Report from the 8th International Conference on Microwave and High Frequency Heating*, 2006, 695-702.
77. SAE International, "Droplet Impingement and Ice Accretion Computer Codes," ARP5903, 2015.
78. Wright, W. and Rutkowski, A., "Validation Results for LEWICE 2.0," NASA Technical Report NASA/CR-1999-208690, 1999.
79. Hann, R., "UAV Icing: Ice Accretion Experiments and Validation," SAE Technical Paper 2019-01-2037, 2019, https://doi.org/10.4271/2019-01-2037.
80. Pasquale, D.D., Rona, A., and Garrett, S., "A Selective Review of Transition Modelling for CFD," in *39th AIAA Fluid Dynamics Conference*, San Antonio, TX.
81. Habashi, W.G., Aubé, M., Baruzzi, G., Morency, F., Tran, P., and Narramore, J.C., "FENSAP-ICE: A Fully-3D in-Flight Icing Simulation System for Aircraft, Rotorcraft and UAVs," in *24th International Congress of the Aeronautical Sciences, ICAS*, Yokohama, Japan, 2004.
82. Szilder, K. and Lozowski, E.P., "Simulation of Airfoil Icing with a Novel Morphogenetic Model," *Journal of Aerospace Engineering* 18:102-110, 2005.
83. Anderson, D.N., "Manual of Scaling Methods," NASA Technical Report NASA/CR-2004-212875, 2004.
84. Bellussi, E., "Use of HISS Test Results for Rotorcraft Civil Certification," SAE Technical Paper 2015-01-2128, 2015, https://doi.org/10.4271/2015-01-2128.
85. ICAO, "Unmanned Aircraft Systems (UAS)," ICAO Technical Report CIR328, 2011.
86. Canadian Aviation Regulations, "Remotely Piloted Aircraft Systems," SOR/96-433, 2019.
87. European Commission, "Commission Implementing Regulation (EU) 2019/947 of 24 May 2019 on the Rules and Procedures for the Operation of Unmanned Aircraft," C/2019/3824, 2019.
88. North Atlantic Treaty Organization (NATO), "STANAG 4671 - Unmanned Aircraft Systems Airworthiness Requirements (USAR), Third Edition," Standardization Agreement, 2019.

Contact Information

EDGEresearch@sae.org.

The authors of this document together with the SAE Team responsible for its creation join in expressing our deepest appreciation to all of the individuals who contributed.

www.ingramcontent.com/pod-product-compliance
Lightning Source LLC
LaVergne TN
LVHW070912210926
848482LV00017B/1059
9781468601695